ASE Test Preparation Series

Transit Bus Test

Heating, Ventilation, and Air Conditioning (HVAC) Systems

(Test H7)

1st Edition

DELMAR
CENGAGE Learning™

Australia • Brazil • Japan • Korea • Mexico • Singapore • Spain • United Kingdom • United States

DELMAR
CENGAGE Learning™

ASE Test Preparation Series: Transit Bus Test: Heating, Ventilation, and Air Conditioning (HVAC) Systems (Test H7), First Edition

Vice President, Technology Professional Business Unit: Gregory L. Clayton

Director, Professional Automotive and Trucking Learning Solutions: Kristen L. Davis

Product Manager: Kimberley Blakey

Editorial Assistant: Jason Yager

Director of Marketing: Beth A. Lutz

Marketing Coordinator: Jennifer Stall

Production Director: Patty Stephan

Production Manager: Andrew Crouth

Content Project Manager: Kara A. DiCaterino

Art Director: Robert Plante

Cover Design: Michael Egan

ISBN-13: 978-1-4180-6571-3

ISBN-10: 1-4180-6571-4

Delmar
Executive Woods
5 Maxwell Drive
Clifton Park, NY 12065
USA

Cengage Learning is a leading provider of customized learning solutions with office locations around the globe, including Singapore, the United Kingdom, Australia, Mexico, Brazil, and Japan. Locate your local office at **www.cengage.com/global**

Cengage Learning products are represented in Canada by Nelson Education, Ltd.

To learn more about Delmar, visit **www.cengage.com/delmar**

Purchase any of our products at your local bookstore or at our preferred online store **www.cengagebrain.com**

Printed in the United States of America
3 4 5 6 7 16 15 14 13 12

Contents

Section 5 Sample Test for Practice

Section 6 Additional Test Questions for Practice

Section 7 Appendices

Preface

Delmar, Cengage Learning is very pleased that you have chosen our ASE Test Preparation Series to prepare yourself for the Transit Bus Heating, Ventilation and Air Conditioning (H7) ASE Examination. This guide is designed to introduce you to the Task List for the HVAC Systems test you are preparing to take, give you an understanding of what you are expected to be able to do in each task, and take you through sample test questions formatted in the same way the ASE texts are structured.

If you have a basic working knowledge of the discipline you are testing for, you will find the Delmar's ASE Test Preparation Guide to be an excellent way to understand the "must know" items to pass the test. This book is not a textbook. Its objective is to prepare the technician who has the requisite experience and schooling to challenge ASE testing. It cannot replace the hands-on experience or the theoretical knowledge required by ASE to master vehicle repair technology. If you are unable to understand more than a few of the questions and their explanations in this book, it could be that you require either more shop-floor experience or further study.

The book begins with an item by item-by-item overview of the ASE Task List with explanations of the minimum knowledge you must possess to answer questions related to the task. Following that there are 2 sets of sample questions followed by an answer key to each test and an explanation of the answers to each question. A few of the questions are not strictly ASE format but were included because they help teach a critical concept that will appear on the test. We suggest that you read the complete Task List Overview before taking the first sample test. After taking the first test, score yourself and read the explanation to any questions that you were not sure about, including the questions you answered correctly. Each test question has a reference back to the related task or tasks that it covers. This will help you to go back and read over any area of the task list that you are having trouble with. Once you are satisfied that you have all of your questions answered from the first sample test, take the additional test and check it. If you pass these tests, you will do well on the ASE test.

Our Commitment to Excellence

Delmar, Cengage Learning has sought out the best technicians in the country to help with the development of this 1st edition of the Transit Bus Heating, Ventilation and Air Conditioning (HVAC) System (H7) ASE Test Preparation Guide.

Thank you for choosing Delmar's ASE Test Preparation Series. All of the writers, editors, and Delmar staff have worked very hard to make this series second to none. We know you are going to find this book easy to use. It is our ongoing objective to improve our products at Delmar by responding to feedback. If you have any questions concerning the books in this series, email us at: autoexpert@trainingbay.com.

1 The History and Purpose of ASE

ASE began as the National Institute for Automotive Service Excellence (NIASE). It was founded as a nonprofit independent entity in 1972 by a group of industry leaders with the single goal of providing a means for consumers to distinguish between incompetent and competent mechanics. ASE accomplishes this goal by testing and certifying repair and service professionals. From this beginning it has evolved to be known simply as ASE (Automotive Service Excellence) and currently offers more than 40 certification exams in automotive, medium/heavy-duty truck, collision, engine machinist, school bus, parts specialist, automobile service consultants, and other industry-related areas.

Today, more than 400,000 professionals hold current ASE certifications. These technicians are employed by new car and truck dealerships, independent garages, fleets, service stations, franchised service facilities, and more. ASE continues its mission by also providing information to help consumers identify repair facilities that employ certified professionals through its Blue Seal of Excellence Recognition Program. Shops that employ a minimum of 75 percent ASE-certified repair technicians and meet other criteria can apply for and receive the Blue Seal of Excellence Recognition.

ASE recognized that educational programs serving the service and repair industry also needed a way to be recognized as having the faculty, facilities, and equipment to provide a quality education to students wanting to become service professionals. Though the combined efforts of the ASE and industry and education leaders, the nonprofit National Automotive Technicians Education Foundation (NATEF) was created to evaluate and recognize training programs. Today more than 2000 programs are ASE certified under the standards set by the service industry. In addition, ASE/NATEF offers certification of industry (factory) training programs through Continuing Automotive Service Education (CASE). CASE recognizes training provided by replacement parts manufacturers as well as vehicle manufacturers.

ASE certification testing is administered by the American College Testing (ACT) service. Strict standards of security and supervision at the test centers ensure that the technician who holds the certification earned it. In addition, ASE certification requires that technicians passing the test demonstrate that they have two years of work experience in the field before they can be certified. Industry experts who are actually working in the field being tested develop test questions. More detailed information on how the test is developed and administered is provided in the next section. Paper-and-pencil tests are administered twice a year at more than 750 locations in the United States. Computer Based Testing (CBT) is now also available with the benefit of instant test results at certain established test centers. ASE certification is valid for 5 years and can be recertified by retesting.

So that consumers can recognize certified technicians, ASE issues a jacket patch, certificate, and wallet card to certified technicians and makes signs available to facilities that employ ASE-certified technicians.

You can contact ASE at:

National Institute for Automotive Service Excellence
101 Blue Seal Drive S.E.
Suite 101
Leesburg, VA 20175
Telephone 703-669-6600
FAX 703-669-6123
www.ase.com

WE SUPPORT PROFESSIONAL CERTIFICATION THROUGH THE National Institute for AUTOMOTIVE SERVICE EXCELLENCE

2 Take and Pass Every ASE Test

Participating in an Automotive Service Excellence (ASE) voluntary certification program gives you a chance to show your customers that you have the "know-how" needed to work on today's modern vehicles. ASE certification tests allow you to compare your skills and knowledge to the automotive service industry's standards for each specialty area.

If you are the "average" automotive technician taking this test, you are in your mid-30s and have not attended school for about 15 years. That means you probably have not taken a test in many years. Some technicians, on the other hand, have attended college or taken postsecondary education courses and may be more familiar with taking tests and with test-taking strategies. There is, however, a difference in the ASE test you are preparing to take and the educational tests you may be accustomed to.

How Are the Tests Administered?

ASE tests are administered at more than 750 test sites in local communities. Paper-and-pencil tests are the type most widely available to technicians. Each tester is given a booklet containing questions with charts and diagrams where required. You can mark in this test booklet but no information entered in it is scored. Answers are recorded on a separate answer sheet. You enter your answers, in number 2 pencil only. ASE recommends you bring 4 sharpened number 2 pencils with erasers. Answer choices are recorded by marking (bubble in) a block by the question number on the answer sheet. These sheets are scanned electronically and the answers tabulated. For test security booklets include randomly generated questions. Your answer key must be matched to the proper booklet so it is important to enter the booklet serial number correctly on the answer sheet. All instructions are printed on the test materials and should be followed carefully.

ASE has introduced Computer Based Testing (CBT) at some locations. While test content is the same for both testing methods, CBT tests have some unique requirements and advantages. If you are considering the CBT tests, visit the ASE web page (www.ASE.com), and review the conditions and requirements for this type of test. A test demonstrator page allows you to experience this type of test before you register. Some technicians find that this style of testing provides an advantage while others find operating the computer a distraction. One significant benefit of CBT is the availability of instant results. You can receive your test results before you leave the test center. CBT testing offers an increased degree of flexibility in scheduling. Testing costs for CBT tests are slightly higher than paper-and-pencil tests and the number of testing sites is limited. First-time test takers may be more comfortable with the paper-and-pencil tests but each technician has an option.

Who Writes the Questions?

Service industry experts in the area being tested write the questions. Each area has its own technical experts. Questions are entirely job related and are designed to test the skills you need to be a successful technician. Theoretical knowledge is important and necessary to answer the questions but ability to apply that knowledge is the basis of ASE test questions.

Each question has its roots in an ASE "item-writing" workshop where service representatives from automobile manufacturers (domestic and import), aftermarket parts and equipment manufacturers, working technicians, and vocational educators meet to share ideas and translate them into test questions. Each test question written by these experts must survive review by all members of the group. Questions are written to address practical application of soft skills and system knowledge experienced by technicians in their day-to-day work.

All questions are pretested and quality-checked on a national sample of technicians. Questions that meet ASE standards of quality and accuracy are included in the scored sections of the tests; the "rejects" are sent back to the drawing board or discarded altogether.

Each certification test is made up of 40—80 multiple-choice questions.

Note: Each test could contain additional questions that are included for statistical research purposes only. Your answers to these questions will not affect your score, but since you do not know which ones they are, you should answer all questions in the test. The 5-year Recertification Test covers the same content areas as those listed above. However, the number of questions in each content area of the Recertification Test will be reduced by about one half.

Objective Tests

A test is called an objective test if the same standards and conditions apply to everyone taking the test and there is only one correct answer to each question.

Objective tests primarily measure your ability to recall information. A well-designed objective test can also test your ability to understand, analyze, interpret, and apply your knowledge. Objective tests include true-false, multiple choice, fill in the blank, and matching questions. ASE's tests consist exclusively of 4-part multiple-choice objective questions.

The following are some strategies that may be applied to your tests.

Before beginning to take an objective test, quickly look over the test to determine the number of questions, but do not try to read through all of the questions. In an ASE test there are usually 40—80 questions, depending on the subject. Read through each question before marking your answer. Answer the questions in the order they appear on the test. Leave questions that you are not sure of blank and move on to the next question. You can return to the unanswered questions after you have finished the others. They may be easier to answer after your mind has had additional time to consider them on a subconscious level. In addition, you might find information in other questions that will help you recall the answers to some of them.

Do not be obsessed by the apparent pattern of responses. For example, do not be influenced by a pattern like **D, C, B, A, D, C, B, A** on an ASE test.

There is also a lot of folk wisdom about taking objective tests. For example, some people advise that you avoid response options that use certain words such as *all*, *none*, *always*, *never*, *must*, and *only*, to name a few. This, they claim, is because nothing in life is exclusive. They would advise that you choose response options that use words that allow for some exception, such as *sometimes*, *frequently*, *rarely*, *often*, *usually*, *seldom*, and *normally*. Some people would also advise that you avoid the first and last option (A and D) because they feel test writers are more comfortable if they put the correct answer in the middle (B and C) of the choices. Another recommendation often offered is to select the option that is either shorter or longer than the other three choices because it is more likely to be correct. Some would advise that you never change an answer since your first intuition is usually correct.

Although there may be a grain of truth in this folk wisdom, it is not wise to make your selections based on the length or order of the answer choices. ASE test writers try to ensure that there are just as many **A** answers as there are **B** answers, just as many **D** answers as **C** answers. As a matter of fact, ASE tries to balance the answers at about 25 percent per choice **A**, **B**, **C**, and **D**. There is no intention to use "tricky" words, such as those outlined above. Put no credence in the opposing words "sometimes" and "never," for example.

Multiple-choice tests are sometimes challenging because there are often several choices that may seem possible, and it may be difficult to decide on the correct one. The best strategy, in this case, is to first determine the correct answer before looking at the options. If you see the answer you decided on, you should still examine the options to make sure that none seem more correct than yours. If you do not know or are not sure of the answer, read each option very carefully and try to eliminate those options that you know to be wrong. That way, you can often arrive at the correct choice through a process of elimination.

If you have gone through all of the test and you still do not know the answer to some of the questions, *then guess*. Yes, guess. You then have at least a 25 percent chance of being correct. If you leave the question blank, you have no chance. Your score is based on the number of questions answered correctly.

Preparing for the Exam

We have included many sample and practice questions in this guide simply to help you learn what you know and what you do not know. We recommend that you work your way through each question in this book. Before doing this, carefully look through Section 3; it contains a description and explanation of the type of questions you'll find in an ASE exam.

Once you understand what the questions will look like, move to the sample test. Answer one of the sample questions (Section 5) then read the explanation (Section 7) to the answer for that question. If you don't feel you understand the reasoning for the correct answer, go back and read the overview (Section 4) for the task that is related to that question. If you still don't feel you have a solid understanding of the material, identify a good source of information on the topic, such as a textbook, and do some more studying.

After you have completed all of the sample test items and reviewed your answers, move to the additional questions (Section 6). This time answer the questions as if you were taking the actual test. Do not use any reference or allow any interruptions so that you can get a feel for how you will do on an actual test. Once you have answered all of the questions, grade your results using the answer key in Section 7. For every question that you answered incorrectly, study the explanations to the answers and/or the overview of the related task areas. Try to determine the cause for missing the answer. The easiest thing to correct is learning the correct technical content. The hardest things to correct are behaviors that lead you to a wrong conclusion. If you knew the information but still got it wrong, there is a behavior problem that will need to be corrected. For example, you may be reading too quickly and skipping over words that affect your reasoning. If you can identify what you did that caused you to answer the question incorrectly, you can eliminate that cause and improve your score. Here are some basic guidelines to follow while preparing for the exam:

- Focus your studies on those areas in which you are weak.
- Be honest with yourself while determining if you understand something.
- Study often but in short periods of time.
- Remove yourself from all distractions while studying.
- Keep in mind the goal of studying is not just to pass the exam, the real goal is to learn!
- Prepare physically by getting a good night's rest before the test and eat meals that provide energy but do not cause discomfort.
- Arrive early to the test site to avoid long waits as test candidates check in and to allow all of the time available for your tests.

During the Test

On paper-and-pencil tests you will be placing your answers on a sheet where you will be required to blacken (bubble) in your answer choice. Stray marks or incomplete erasures may be picked up as an answer by the electronic reader so be sure only your answers end up on the sheet. One of the biggest problems an adult faces in test taking is in placing an answer in the correct spot on a bubble sheet. Make certain that you mark your answer for, say, question 21, in the space on the bubble sheet designated for the answer for question 21. A correct response in the wrong line will probably result in two questions being marked wrong, one with two answers (which could include a correct answer but will be scored wrong), and the other with no answer. Remember, the answer sheet on a written test is machine scored and can only "read" what you have bubbled in.

If you finish answering all of the questions on a test and have remaining time, go back and review the answers to those questions that you were not sure of. You can often catch careless errors by using the remaining time to review your answers. Carefully check your answer sheet for blank answer blocks or missing information.

At practically every test, some technicians will invariably finish ahead of time and turn in their papers long before the final call. Some technicians may be doing recertification tests and others may be taking fewer tests than you. Do not let them distract or intimidate you.

It is not wise to use less than the total amount of time that you are allotted for a test. If there are any doubts, take the time for review. Any product can usually be made better with some additional effort. A test is no exception. It is not necessary to turn in your test paper until you are told to do so.

Testing Time Length

An ASE written test session is 4 hours. You may attempt from 1 to a maximum of 4 tests in one session. It is recommended, however that no more than a total of 225 questions be attempted at any test session. This will allow for just over 1 minute for each question.

Visitors are not permitted at any time. If you want to leave the test room, for any reason, you must first ask permission. If you finish your test early and want to leave, you are permitted to do so only during specified dismissal periods. You are not permitted back in the test room once you leave after completing your test.

You should monitor your progress and set an arbitrary limit as to how much time you will need for each question. This should be based on the number of questions you are attempting. Wear a watch because some facilities may not have a clock visible to all areas of the room.

CBTs are allotted a testing time according to the number of questions ranging from 1/2 hour to 1 1/2 hours. Advanced-level tests are allowed 2 hours. This time is by appointment so arrive on time to ensure that you have all of the time allocated. If you arrive late for a CBT test appointment, you will only have the amount of time remaining on your appointment.

Your Test Results!

You can gain a better perspective about tests if you know and understand how they are scored. ASE's tests are scored by American College Testing (ACT) service, a nonpartial, unbiased organization having no vested interest in ASE or in the automotive industry.

Each question carries the same weight as any other question. For example, if there are 50 questions, each is worth 2 percent of the total score. The passing grade is 70 percent. That means you must correctly answer 35 of the 50 questions to pass the test.

The test results can tell you:

- where your knowledge equals or exceeds that needed for competent performance, or
- where you might need more preparation.

Your ASE test score report is divided into content areas and will show the number of questions in each content area and how many of your answers were correct. These numbers provide information about your performance in each area of the test. However, because there may be a different number of questions in each content area of the test, a high percentage of correct answers in an area with few questions may not offset a low percentage in an area with many questions.

One does not "fail" an ASE test. The technician who does not pass is simply told "More Preparation Needed." Though large differences in percentages may indicate problem areas, it is important to consider how many questions were asked in each area. Since each test evaluates all phases of the work involved in a service specialty, you should be prepared in each area. A low score in one area could keep you from passing an entire test. There is no such score as average. You cannot determine your overall test score by adding the percentages given for each task area and dividing by the number of areas. Test scoring does not work that way because generally there are not the same number of questions in each task area. A task area with 20 questions, for example, counts more toward your total score than a task area with 10 questions.

Your test report should give you a good picture of your results and a better understanding of your strengths and weaknesses for each task area.

If you do not pass the test, you may take it again at any time it is scheduled to be administered. You are the only one who will receive your test score. Test scores will not be given over the telephone by ASE nor will they be released to anyone without your written permission.

3 Types of Questions on an ASE Exam

ASE certification tests are often thought of as being tricky. They may seem tricky if you do not completely understand what is being asked. The following examples will help you recognize certain types of ASE questions and avoid common errors.

Paper-and-pencil tests and Computer Based Test questions are identical in content and difficulty. Most initial certification tests are made up of 40–80 multiple-choice questions. Multiple-choice questions are an efficient way to test knowledge. To answer them correctly, you must think about each choice as a possibility, and then choose the one that best answers the question. To do this, read each word of the question carefully. Do not assume you know what the question is about until you have finished reading it.

About 10 percent of the questions on an actual ASE exam will use an illustration. These drawings contain the information needed to answer the question correctly. The illustration must be studied carefully before attempting to answer the question. Often, technicians look at the possible answers then try to match up the answers with the drawing. Always do the opposite; match the drawing to the answers. When the illustration is showing an electrical schematic or another system in detail, look over the system and try to figure out how the system works before you look at the question and the possible answers.

Multiple-Choice Questions

The most common type of question used on ASE Tests is the multiple-choice test. This type of question contains 3 "distracters" (wrong answers) and 1 "key" (correct answer). When the questions are written, effort is made to make the distracters plausible to draw an inexperienced technician to one of them. This type of question gives a clear indication of the technician's knowledge. Using multiple criteria including cross-sections by age, race, and other background information, ASE is able to guarantee that a question does not bias for or against any particular group. A question that shows bias toward any particular group is discarded.

If you encounter a question that you are unsure of, reverse engineer it by eliminating the items that it cannot be. For example:

When an expansion valve is properly installed, where should the thermal bulb and capillary tube be positioned?

A. Fastened to the condenser fins using epoxy

B. Located in the accumulator

C. Held in contact with the evaporator outlet using insulating tape

D. As an integral part of the orifice tube assembly (B3.6)

Answer C is correct. The thermal bulb and capillary tube are part of the thermal expansion valve located near the evaporator inside of the duct housing. The thermal bulb and capillary tube sense the temperature of the outlet line of the evaporator core. When the thermal bulb senses warm temperatures, the TXV opens to allow more refrigerant into the evaporator core. When the thermal bulb senses cold temperatures, the TXV closes to restrict refrigerant flow into the evaporator.

Answer A is incorrect. The expansion valve is located near the evaporator, not the condenser.

Answer B is incorrect. An accumulator is used with a CCOT system that does use a thermal bulb or capillary tube.

Answer D is incorrect. A CCOT system does use a thermal bulb or capillary tube.

EXCEPT Questions

Another type of question used on ASE tests has answers that are all correct except one. The correct answer for this type of question is the answer that is wrong. The word "EXCEPT" will always be in capital letters. You must identify which of the choices is the wrong answer. If you read quickly through the question, you may overlook what the question is asking and answer the question with the first correct statement. This will make your answer wrong. An example of this type of question and the analysis is as follows:

When servicing the expansion valve, a technician can perform the following EXCEPT:

A. Adjust the expansion valve using a torque wrench.

B. Adjust the expansion valve using an Allen wrench.

C. Adjust the expansion valve using a screwdriver.

D. The expansion valve cannot be adjusted. (B3.6)

Answer D is correct. The expansion valve is an adjustable device.

Answer A is incorrect. You can adjust an expansion valve.

Answer B is incorrect. You can adjust an expansion valve.

Answer C is incorrect. You can adjust an expansion valve.

Technician A, Technician B Questions

The type of question that is most popularly associated with an ASE test is the "Technician A says . . . Technician B says . . . Who is correct?" type. In this type of question, you must identify the correct statement or statements. To answer this type of question correctly, carefully read each technician's statement and judge it on its own merit to determine if the statement is true.

Sometimes this type of question begins with a statement about some analysis or repair procedure. This is often referred to as the stem of the question and provides the setup or background information required to understand the conditions on which the question is based. This is followed by two statements about the cause of the concern, proper inspection, identification, or repair choices. You are asked whether the first statement, the second statement, both statements, or neither statement is correct. Analyzing this type of question is a little easier than the other types because there are only two ideas to consider although there are still four choices for an answer.

Technician A, Technician B questions are really double true or false questions. The best way to analyze this kind of question is to consider each technician's statement separately. Ask yourself, is A true or false? Is B true or false? Then select your answer from the four choices. Remember, an ASE Technician A, Technician B question will never have Technician A and B directly disagreeing with each other. That is why you must evaluate each statement independently.

An example of this type of question and the answer for it follows.

The A/C condenser has several bent fins and a moderate accumulation of dead insects. Technician A says that the condenser should be replaced or it will cause the high-pressure relief valve to trip. Technician B says that the condenser fins should be straightened and the dead insects removed to optimize A/C system performance. Who is correct?

A. A only

B. B only

C. Both A and B

D. Neither A nor B (B3.3)

Answer B is correct. Only technician B is correct. Several bent fins and a moderate accumulation of dead insects will restrict airflow through the condenser enough to cause poor cooling and higher-than-normal system pressures. It is a good practice to check for condenser restrictions while performing a good preventative maintenance inspection.

Answer A is incorrect. The condenser should not be replaced, the fins should be straightened and then it should be cleaned.

Answer C is incorrect. Only Technician B is correct.

Answer D is incorrect. Only Technician B is correct.

Most-Likely Questions

Most-Likely questions are somewhat difficult because only one choice is correct while the other three choices are nearly correct. An example of a Most-Likely-cause question is as follows:

An electric oil pressure gauge indicates zero pressure, but the engine is well lubricated. During an HVAC performance test, the technician hears the A/C compressor clutch slip briefly upon engagement. The Most-Likely cause is:

A. a worn out compressor clutch coil.

B. a defective A/C compressor clutch relay.

C. the compressor clutch air gap is too large.

D. the compressor clutch bearing is worn. (A2 and A4)

Answer C is correct. If the air gap is too large, the clutch may slip briefly upon engagement.

Answer A is incorrect. If the compressor clutch coil were defective, then the clutch would not engage at all.

Answer B is incorrect. A defective relay will prevent the clutch from engaging.

Answer D is incorrect. A worn clutch bearing may cause a loud clutch engagement but not a slipping clutch.

LEAST-Likely Questions

Notice that in Most-Likely questions not all the letters are capitalized. This is not so with LEAST-Likely questions. For this type of question, look for the choice that would be the LEAST-Likely cause of the described situation. Read the entire question carefully before choosing your answer. Avoid relating questions to those unusual situations that you may have encountered and answer based on the technical and mechanical possibilities.

An example is as follows:

When attempting to verify a leaking evaporator core, the technician is LEAST-Likely to sense refrigerant with the detector probe:

A. at the evaporator housing drain.

B. over the expansion valve.

C. at the overhead vents.

D. at the defroster vents. (B3.8)

Answer D is correct. The defroster vent is not part of the A/C system. The evaporator drain is the Most-Likely place to detect refrigerant leaking from the evaporator core.

Answer A is incorrect. Refrigerant leaking from the evaporator core could possibly be detected in the evaporator drains

Answer B is incorrect. Refrigerant leaks may be detectable above the expansion valve.

Answer C is incorrect. Refrigerant leaking from the evaporator core could possibly be detected at the overhead vents.

Summary

There are no four-part multiple-choice ASE questions having "none of the above" or "all of the above" choices. ASE does not use other types of questions, such as fill-in-the-blank, completion, true-false, word-matching, or essay. ASE does not require you to draw diagrams or sketches. If a formula or chart is required to answer a question, it is provided for you. There are no ASE questions that require you to use a pocket calculator.

4 Overview of Task List

Heating, Ventilation, and Air Conditioning (HVAC) Systems (Test H7)

The following section includes the task areas and task lists for this test and a written overview of the topics covered in the test.

The task list describes the actual work you should be able to do as a technician and on which you will be tested by the ASE. This is your key to the test and you should review this section carefully. We have based our sample test and additional questions upon these tasks. The overview section will also support your understanding of the task list. ASE advises that the questions on the test may not equal the number of tasks listed; the task lists outline the specific tasks that ASE expects you to know how to perform and which you will be tested on.

At the end of each question in the Sample Test and Additional Questions Sections, a letter and number are presented as a reference back to the related task for additional study. Note the following example: A1.

A. Heating, Ventilation, and Air Conditioning Systems Diagnosis, Service, and Repair (9 Questions)

Task A1 Verify the need for service or repair of HVAC systems based on operating noises; determine appropriate action.

Example:

1. The driver of a bus states that the air conditioning has not worked well since the bus was repaired. Technician A states that prior service history is not of any interest to a technician. Technician B states that the refrigerant should be recovered to see if the system has a low charge. Who is correct?
   A. A only
   B. B only
   C. Both A and B
   D. Neither A nor B (A1)

Question #1

Answer D is correct. Neither technician is correct. The first step of any diagnosis is to verify the complaint and to gather as much background information about the vehicle as possible. This will help the technician get to the root cause of the problem and correct it.

Answer A is incorrect. Technician A is incorrect. Checking the prior service history on a vehicle often leads the technician to the problem more quickly.

Answer B is incorrect. Technician B is incorrect. A technician should never recover the refrigerant as the first step of repair. The problem should be verified, researched, and diagnosed before any repair work is done. It is also very important to identify the refrigerant before recovering it into your equipment.

Answer C is incorrect. Neither technician is correct.

Task List and Overview

A. Heating, Ventilation, and Air Conditioning Systems Diagnosis, Service, and Repair (9 Questions)

Task A1 Verify the need for service or repair of HVAC systems based on operating noises; determine appropriate action.

The service technician must be aware of normal HVAC system operating noises in order to determine whether a system requires service. Normal noises include the sounds of A/C compressor clutch engagement, the blower motor, and pressure equalization after the vehicle is shut down. Noises that could indicate the need for service include a growling sound from the water pump or A/C compressor. A loose, dry, or worn A/C compressor belt will cause a squealing noise. This noise will be worse during acceleration. Worn or dry blower motor bearings may cause a squealing noise when the blower is running; this noise may occur when the engine first starts after sitting overnight. A loose or worn clutch hub or loose compressor mounting bolt will also cause a rattling noise from the compressor.

If liquid refrigerant enters the compressor, a thumping, banging noise will result. Heavy knocking compressor noises are caused by refrigerant system blockage, incorrect pressures, or internal damage. A worn compressor pulley bearing will cause a growling noise with the compressor disengaged. If the growling noise only occurs when the system engages the clutch, a worn clutch armature or pulley could be the fault.

Task A2 Verify the need for service or repair of HVAC systems based on sight, odor, and touch conditions; determine appropriate action.

The service technician must be aware of abnormal conditions in the HVAC system in order to determine the need for system service. If the driver complains of high or low temperatures inside the bus, this is cause for a system performance test. Abnormal conditions include the smell of antifreeze inside the bus, a fogged windshield, ice buildup on A/C components, and oil or dirt buildup on A/C fittings.

Task A3 Identify HVAC system components and refrigerant type (R-22, R-134a, R-407c); conduct performance tests; determine appropriate action.

First, the technician must know if the system operates on R-134a, R-22, or R-407c refrigerant. Most major R-134a components use a light-blue label to indicate an R-134a system. Systems use Schrader-type service valves, and R-134a systems may use metric threads and quick connect/disconnect service valves.

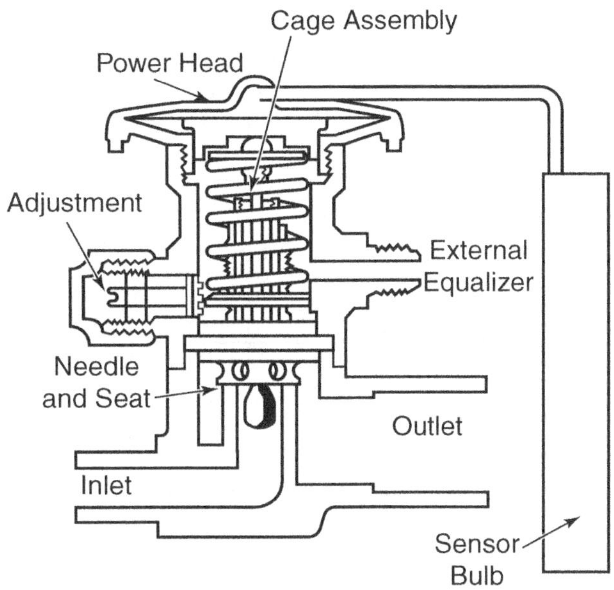

Next, you need to identify the type of expansion device in the evaporator inlet line. Most A/C systems use a thermal expansion valve (TXV), as shown in the figure, in the evaporator inlet line to control refrigerant flow into the evaporator. The system has high-pressure liquid on the inlet side of the expansion device and low-pressure liquid on the outlet side.

The bus industry uses three types of A/C methods to control temperature: mechanical, electronic thermostat, or microprocessor control. Mechanical systems have a thermostat that uses a capillary tube to a bellows that operates a switch to turn the compressor on or off. A rotary temp switch is used to control the in-bus temperature manually. In an electronic thermostat system, bus temperature is controlled by solid state components, but in the microprocessor-controlled system, most subsystems are computer controlled.

Manual systems rely on the driver to select the temperature, mode, and blower speed. Electronic thermostat temperature control (ETC) systems regulate only the temperature of the output air and rely on the driver to select the desired mode and blower speed. A microprocessor (computer) controls a fully automatic temperature control system. These systems use the input from various sensors in the vehicle to adjust the interior temperature using an appropriate blower speed or coolant to the heater core for reheat cycles, or cycling the compressor clutch.

An HVAC performance test should include operation of the system in all modes and at a variety of temperatures and blower speeds. A small, pocket thermometer that is known to be accurate should be used to verify that the output air temperatures match the temperature settings.

Task A4 Use a data reader/computer to determine fault codes and perform system tests; check and adjust system parameters; clear fault codes.

Computer-controlled HVAC systems use fault codes to determine problem areas. Thermo King Corporation uses two systems, IntelligAIRE 1 and IntelligAire 11. IntelligAIRE 1 uses a dedicated reader to set up, calibrate, read and clear codes. The reader can also be used to perform diagnostic tests. See example for codes.

Software Codes

Code	Code Description
00	no alarm
01*	evaporator coil freeze for 3 minutes (program 15 only)
02	evaporator coil low temperature sensor (shorted, open or backwards)
03	return air tempeture sensor (shortened, open or backwards)
04	AUX sensor 1 (Only if installed)
05	compressor discharge sensor (shorted, open or backwards)
06	compressor discharge high tempeture sensor (Open)
07	low heating capacity (rev 5.5 only)
08	compressor discharge high tempeture warning
09*	evaporator coil high temperature for 5 minutes (program 15)
10	high discharge pressure cutout
11	low suction pressure cutout
12	compressor discharge high tempature shutdown
13**	failure of condenser motor drive to start after 5 attempts
14**	failure of evaporator motor drive to start after 5 attempts

B. A/C System and Component Diagnosis, Service, and Repair (27 Questions)

1. A/C System — General (13 Questions)

Task B1.1 Diagnose the cause of A/C system temperature control problems; determine needed repairs.

To check for causes of temperature control problems, begin by checking for compressor clutch engagement when A/C is selected. You should hear the clutch engage and notice a change in engine RPM and sound. If you are unable to tell by this method, watch the compressor clutch while an assistant turns the A/C on. If the clutch fails to engage, check the wiring schematic for the system to identify the power and ground sources and any compressor control devices used. Follow the manufacturer's procedures to pinpoint the source of the failure. If the compressor engages, a quick check to see if the system is operating can be made by comparing the compressor suction and discharge temperatures. The suction side should be cool to the touch and the discharge should be hot. Perform further testing by attaching a manifold gauge set and doing a performance test. Compare to the manufacturer's specifications. After verifying that refrigerant temperatures and pressures are acceptable, check for air temperature control system problems.

The temperature of A/C system output air is generally controlled by one of two ways. In cycling clutch systems, air is cooled by the evaporator core. Ultimately, the output air temperature control occurs by regulating the amount of time the compressor is running. In A/C systems with reheat, temperature control is achieved by opening and closing the water control valve. When the temperature selector is in the full cold position, the water control valve totally blocks all hot coolant from entering the heater core. When the temperature selector is moved toward the hot setting, the water control valve opens to allow hot coolant into the heater core, resulting in warmer air being delivered to the bus. This type of system typically has the evaporator core and the heater core stacked next to each other. A/C system outlet air temperature is affected by outside air temperature and humidity, engine coolant temperature, airflow through the condenser and evaporator, and level of refrigerant charge. Outlet air temperature may also be affected by mechanical or electrical failure of system components.

Task B1.2 Identify A/C refrigerant type; check for contamination; determine appropriate action.

The easiest way to identify the type of refrigerant that is used in a given A/C system is to observe the labels or the unit identification tag. R-134a systems may use quick-disconnect fittings. You cannot vent R-22 or R-134a to the atmosphere. A refrigerant identifier machine may be connected to all types of A/C systems so as to determine refrigerant type and level of contamination. (*Note*: Recovery machines can be damaged by recovery of chemicals not compatible with their seals and valves.)

Task B1.3 Diagnose the cause of A/C system problems indicated by pressure gauge readings and sight glass/moisture indicator conditions (where applicable); determine needed service or repairs.

In a normally operating A/C system with R-134a, the low-side pressure varies between 10 and 45 psi, and the high-side pressure varies between 120 and 300 psi, depending on the ambient temperature and humidity levels. The following table summarizes abnormal A/C system pressures and common causes.

A/C Pressure Diagnosis

LOW-SIDE PRESSURE	HIGH-SIDE PRESSURE	POSSIBLE CAUSES
LOW	LOW	Low refrigerant charge
LOW	LOW	Obstruction in the suction line
LOW	LOW	Clogged orifice tube
LOW	LOW	TXV stuck closed*
LOW	LOW	Restricted line from the condenser to the evaporator*
LOW	HIGH	Restricted evaporator air flow
HIGH	LOW	Internal compressor damage
HIGH	HIGH	Refrigerant overcharge
HIGH	HIGH	Restricted condenser air flow
HIGH	HIGH	High engine coolant temperature
HIGH	HIGH	TXV stuck open
HIGH	HIGH	Air or moisture in the refrigerant

*Stuck closed TXVs or a restricted line from the condenser to the evaporator will cause frosting at the point of restriction.

In A/C systems, a sight glass allows the service technician to make a quick assessment of the system condition. With the A/C compressor clutch engaged, a properly charged system will occasionally show traces of bubbles. A sight glass that appears foamy indicates that the refrigerant charge is low. When the sight glass contains bubbles and/or foam, the refrigerant charge is low and refrigerant vapor has entered the system. Oil streaks appearing in the sight glass indicate that compressor oil is circulating through the system. A cloudy sight glass indicates that the desiccant pack in the dryer has broken down or a large amount of acid is in the system.

A clear sight glass may indicate the proper refrigerant charge; it may also indicate an excessive refrigerant charge or no refrigerant charge. Most manufacturers agree that the following conditions must be present when performance testing an R-134a system:

- temperature control set in high cool position
- high blower speed
- engine speed set at 1500 rpm or peak governed rpm
- monitor low-side and high-side pressure
- monitor the temperature of the air from the duct

Task B1.4 Diagnose the cause of A/C system problems indicated by sight, audible, odor, and touch conditions; determine needed repairs.

To diagnose A/C systems efficiently, the service technician must be observant. Restricted hoses cause a frosting or sudden temperature change at a specific point along the hose. Frost on the receiver/dryer usually indicates an internal restriction in that component. Because the receiver/dryer is located in the high-pressure liquid line between the condenser and evaporator, it should feel warm. Frost formation on the evaporator outlet indicates a flooded evaporator, which is caused by an excessive refrigerant charge or a stuck-open TXV. Unusual noises can often guide the technician to a faulty component. These problems may also cause frost on the compressor suction line.

On systems using an accumulator, it should feel cold because of its close connection to the evaporator. Both the evaporator inlet and outlet should feel cold when operating normally. If the evaporator outlet is warm, the refrigerant charge may be low. High-side refrigerant components should feel hot or warm, and low-side components should feel cool or cold. A plugged drain in the evaporator case causes water to be pushed into the overhead vents.

A thorough visual inspection is always a good first step in diagnosing A/C systems.

Task B1.5 Perform A/C system leak test; determine needed repairs.

Refrigerant systems use four methods to detect leaks: dye check, ultra sonic, bubble solution, or electronic leak detector. To check an A/C system for leaks, the technician must first ensure that the system contains enough refrigerant to allow compressor clutch engagement. If the system is empty, you should first pressure test the system using nitrogen. Then when you are sure the system will hold pressure, install a partial refrigerant charge. When the partial charge is installed, a small amount of dye should also be added. Follow the manufacturer's guidelines for the use of dyes. After running with the dye installed for 15 minutes, the dye will appear at the leak area. Ultraviolet dye is available for installation into the refrigerant and is visible using a black light detector. Electronic detectors provide an audible beeping sound when the probe is placed near the leak source. Since R-22 and R-134a are different chemically, a specific electronic detector or one that tests both systems is used. When checking for leaks, place the leak detector probe directly below each fitting, and each component directly below the evaporator drain. Check the entire system to rule out multiple leaks.

When using an electronic leak detector, calibrate the detector before each use. Other common means of leak detection are a soap/water or bubble solution applied to external areas of the system. If leaks are present under pressure, bubbles will appear. Another method is fluorescent soap; as with the dye, it will change color when it comes into contact with the refrigerant.

Under no circumstances is a flame-type tester to be used. When R-22 comes into contact with a flame, it becomes phosgene gas. Also, all refrigerants will combust at the appropriate temperature.

Task B1.6 Evacuate A/C system using appropriate equipment.

Before you can evacuate a system, you must first recover any remaining refrigerant; it is important to follow the manufacturer's instructions for the specific recovery station used. Never vent refrigerant to the atmosphere; it is illegal and environmentally irresponsible.

When the system is completely empty, connect the manifold gauge center hose to a vacuum pump and micron gauge. Operate this pump with the service valves open and the low-side and high-side gauge valves open. When the micron gauge is reading 1200 microns, run for an additional 30 minutes. Isolate the vacuum pump from the system. If the low-side gauge needle rises slowly and then stops rising, this indicates the presence of moisture in the system. Additional time is needed to evacuate the system. If the gauge reading continues to rise to zero, this indicates a leak in the system. Fix the leak and proceed to evacuate the system to at least 1200 microns. Most manufacturers recommend that the vacuum pump run until 1200 microns to ensure that all moisture is removed from the system.

Task B1.7 Remove contaminants from the A/C system.

If desiccant pack deterioration or catastrophic compressor failure occurs, clean all refrigeration system components internally. Internal cleaning of A/C system components is best accomplished using a suction filter.

Bus manufacturers recommend using a suction filter between the evaporator and the compressor to remove debris.

Task B1.8 Charge A/C system with refrigerant.

The technician must complete the recovery and evacuation procedure before charging a refrigerant system. Modern recovery and charging stations do not require the A/C system to operate during system charging. It is always best to read all of the manufacturer's instructions for the specific charging station used. The original equipment manufacturer (OEM) may recommend high-side (liquid) charging procedures, low-side (vapor) charging procedures, or both when the system is not operating.

Connect the center hose to the proper refrigerant container and open the container valve. With the engine not running and using the high-side (liquid) charging process, open the high-side gauge valve and the low-side valve. Charging is complete when the correct weight of refrigerant has entered the system. Close the high- and low-side valves on the gauge manifold. Now start the engine and run an A/C performance test. If the system is charged with the engine running and the A/C turned on, make sure to charge only through the low side.

Task B1.9 Identify A/C system lubricant type needed for system application.

All R-22 systems use mineral-based refrigeration or alklbenzene oil to lubricate the compressor and prevent internal corrosion of components. Mineral-based oil is not compatible with R-134a systems. R-134a systems use polyolester oil. The POE lubricant is synthetic oil and is not compatible with R-12 systems. Many retrofit kits come with ester oil included. If the technician is unable to extract all of the mineral oil during the retrofit process, then the ester oil can be used instead of the mineral oil.

2. Compressor and Clutch (7 Questions)

Task B2.1 Diagnose A/C system problems that cause protection devices (pressure, thermal, and electronic) to interrupt system operation; determine needed repairs.

A variety of A/C system protection devices can be used in bus HVAC systems. The low-pressure cutout switch will interrupt compressor operation if system pressure drops to the point that a loss of refrigerant charge to the compressor occurs. The high-pressure cutout switch interrupts compressor operation in case of extremely high system pressure. The binary switch combines the function of the low- and high-pressure cutout switches. Some systems have a high-pressure relief valve mounted in the high-side liquid line. This valve opens and relieves system pressure if the pressure exceeds 450 to 550 psi (3100 to 3792 kPa). Condenser airflow restrictions cause these extremely high pressures. Cycling clutch systems use a pressure cycling switch to cycle the compressor off and on, in relation to low-side pressure. This switch is mounted in the compressor suction manifold. This switch opens and turns off the compressor clutch when the refrigerant pressure is below 10 psi (315 kPa).

Task B2.2 Inspect and test A/C system pressure, thermal, and electronic protection devices and connections; replace as required.

A/C system pressure protection switches are normally closed and may be tested by checking for continuity using a DVOM or a circuit tester while the system is at normal pressure and temperature conditions. The high-pressure switch opens if air-conditioning pressure exceeds about 425–435 psig to prevent damage and closes when pressure drops to below 200 psig. The normally closed low-pressure switch, which is usually located on the accumulator, opens when the low-side pressure drops below 10 psig. Compressor head temperature switches, when used, are mounted so that they contact the compressor case; when the switches sense that the temperature is exceeding a certain threshold, they will open, turning the compressor clutch off. A clamping diode is usually placed across the compressor clutch coil to reduce voltage spikes, which are caused when the clutch is cycled. This is to protect other voltage-sensitive devices. If the diode is faulty, it will not prevent the clutch from engaging, but will fail to protect other voltage-sensitive parts of the electrical system.

Task B2.3 Inspect and adjust A/C compressor drives, belts, pulleys, and tensioners; replace as required.

When inspecting an A/C system, it is important that the technician not overlook the A/C compressor drive belts and pulleys. Drive belt edge wear indicates a misaligned or bent pulley. If the belt is loose or bottomed out in the pulley, the belt may slip and cause inadequate cooling. Cracked, glazed, or frayed belts must be replaced. Use a standard drive belt tension gauge to check and adjust belt tension. An improperly adjusted drive belt will wear or fail prematurely. A drive belt adjusted too loosely may slip and cause belt squealing, especially on acceleration with the A/C on and clutch engaged. A drive belt adjusted too tightly may cause internal engine wear or damage to other belt-driven components. Cracked or bent pulleys must be replaced, not repaired.

Task B2.4 Inspect, test, and service A/C compressor clutch components and clutch assembly; replace as required.

The A/C compressor clutch assembly allows the A/C compressor to engage and disengage to modulate system pressures. The components of the compressor clutch assembly are:

- the driven plate, which is keyed to the compressor drive shaft
- the drive plate, which is integral to the drive pulley
- the clutch bearing, which operates all the time
- the clutch coil, which creates the magnetic field that engages the clutch

Defective compressor clutch bearings will make a growling noise with the engine running. Test the compressor clutch by applying power and ground to the appropriate terminals and watching for clutch engagement. You must inspect the pulley and armature plate frictional surfaces for wear and oil contamination. Check the hub bearing for roughness, grease leakage, and looseness. An armature that drags on the pulley face or slips briefly on engagement indicates an improper clutch air gap. Adjust the air gap using the adjustment screws. Check the gap on any clutch service. A technician can perform a voltage drop test while the system is in operation. This is the most practical test method as components are in actual running mode.

Task B2.5 Check and correct A/C compressor lubricant level and condition.

Check the A/C compressor lubricant level and adjust it any time there is evidence of lubricant loss from the system. An excessive amount of oil in a refrigerant system reduces system-cooling efficiency. To check the compressor lubricant level, view the oil level in the sight glass. The oil level should be at ¼ to ½ in the sight glass, with the system running and stabilized.

R-22 systems require a mineral oil or synthetic alklbenzene. R-134a systems with a reciprocating compressor must have a synthetic (POE) oil designation. If the oils used become mixed, compressor damage will result.

Task B2.6 Inspect, test, and replace A/C compressor.

You can diagnose A/C compressor internal damage using a standard A/C gauge set. Low high-side pressure and high low-side pressure on the manifold gauge set may indicate a defective compressor. A faulty compressor bearing will make a growling noise with the engine running and the compressor clutch engaged. Oil dripping from the front of the compressor indicates a faulty front seal. A rattling noise may be caused by loose compressor mounts. A defective pulley bearing will also cause a growling noise with the clutch disengaged. When replacing A/C seals or O-rings, prelubricate them with the proper refrigeration oil.

Task B2.7 Inspect and repair A/C compressor mountings and hardware; replace as required.

Damaged A/C compressor mounts or mounting plates can cause drive belt misalignment, improper drive belt tension, and compressor vibration. Welding can repair cracked mounts and mounting plates; however, care must be taken to align all parts properly, as serpentine belts require alignment of components to within 1 degree.

Task B2.8 Check operation of A/C compressor unloaders; adjust as required

Unloaders are part of the compressor system on some models. Their purpose is to decrease compressor capacity when near set points and on low-heat load conditions.

3. Evaporator, Condenser, and Related Components (7 Questions)

Task B3.1 Adjust A/C system lubricant level after replacement of system components.

Most A/C system components contain refrigeration oil. When replacing the evaporator, condenser, receiver, dryer, or A/C hoses, a slight amount of refrigeration oil could be lost. Fresh refrigeration oil should be added to the compressor per the manufacturer's specifications.

Task B3.2 Inspect and repair A/C system hoses, lines, filters, fittings, and seals; replace as required.

You should check A/C system hoses for damage and leaks during the course of any A/C system maintenance or inspection. Hoses should be replaced if they are cracked, kinked, abraded, or if the fittings show any signs of abuse. Disassemble leaking fittings and replace the O-rings. New O-rings should be lubricated with the appropriate refrigeration oil. Some manufacturers recommend that mineral oil be used to lubricate the O-rings. It is important to use the nitril O-rings on R-134a systems.

A/C system filters and screens are used to prevent particulate (from corrosion, compressor failure, or desiccant breakdown) from circulating through the A/C system and must be replaced if they are clogged, restricted, or damaged. Some systems have a filter in the line between the condenser and the evaporator, and some of these filters contain an orifice tube. You must install this type of filter in the proper direction. Special tools are required to service the spring lock couplings that are used on some hose connections.

Task B3.3 Check air flow through A/C condensers and evaporators; correct as required.

The A/C condenser should be checked for proper airflow at regular intervals. During a normal A/C system inspection, any bent condenser fins should be straightened and any debris should be cleaned from the condenser. Debris in the condenser air passages causes excessive high-side and low-side pressures and reduced cooling. This problem may also cause the high-pressure relief valve to discharge refrigerant.

Task B3.4 Inspect and test A/C system condenser and mountings; replace as required.

If any refrigerant tubes are kinked, cracked, or leaking, the A/C condenser must be replaced. Frost on any of the condenser tubing indicates a refrigerant passage restriction. This condition results in excessive high-side and low-side pressures and inadequate cooling. If you replace the condenser, the mounts should be checked for improper alignment and deformation, which could cause abrasion and fatigue damage.

Task B3.5 Inspect receiver and filter/dryer; replace as required.

Most bus A/C systems use a filter/dryer to ensure an adequate supply of clean, dry refrigerant to the system. If the filter/dryer inlet and outlet pipes have a significant temperature difference, the filter/dryer is restricted. Frost forming on the filter/dryer indicates an internal restriction. Bubbles and foam in the sight glass indicate rust and moisture contamination. Both of these devices contain a block of desiccant designed to absorb and hold traces of moisture from the refrigerant. The filter/dryer is located just upstream of the system expansion device. The dryer must be replaced if the A/C system has remained open to the atmosphere for an extended period of time. Other reasons for replacing the drying device are evidence of moisture or corrosion in the system, or if catastrophic compressor failure has occurred.

Task B3.6 Inspect and test refrigerant solenoid, expansion valve(s), and evaporator pressure regulators; check placement of thermal bulb (capillary tube); replace as required.

One type of A/C system expansion device is the thermal expansion valve. The thermal expansion valve senses evaporator temperature using a capillary tube connected to a thermal bulb. As the fluid inside the thermal bulb expands, the orifice in the expansion valve opens to increase refrigerant flow through the evaporator. If the evaporator core temperature drops to near the freezing point, the fluid in the thermal bulb contracts and the expansion valve closes to restrict refrigerant flow. Different designs place this thermal bulb either embedded in the evaporator fins or affixed to the evaporator outlet with insulating tape.

In some systems, the expansion valve is housed in a combination valve or an "H" valve. In these systems, internal mechanisms monitor evaporator inlet and outlet temperatures and pressures and adjust the valve opening accordingly. A technician can diagnose a faulty expansion valve using a set of A/C pressure gauges and consulting a diagnostic chart.

Some A/C systems use a thermostatic switch to monitor evaporator temperature. If the temperature gets too cold, this switch opens to interrupt power to the A/C compressor clutch. These switches are used as de-icing devices in the A/C system.

Task B3.7 Inspect and test evaporator core; replace as required.

Cycling clutch-type A/C systems often use an expansion valve as a metering device. The expansion valve is located at the evaporator inlet and contains a fine screen to prevent the circulation of particulate through the evaporator core and back to the compressor. The expansion valve should be replaced if the screen is restricted or corroded, or in case of desiccant bag breakdown or catastrophic compressor failure.

A restricted expansion valve may cause lower-than-specified low-side pressure, frosting of the expansion valve, and inadequate cooling from the evaporator. If these conditions appear, place a shop towel soaked in hot water around the expansion valve. If the low-side pressure increases, there is moisture freezing in the expansion valve. To rectify this condition, you must recover, evacuate, and recharge the system. If the hot shop towel did not increase the low-side pressure, clean or replace the expansion valve.

Task B3.8 Inspect, clean, and repair evaporator housing and water drain; inspect and service or replace evaporator air filter.

The evaporator case or housing contains the evaporator core, the heater core, and the evaporator core drain. A clogged evaporator core drain will cause water to spray from the overhead vents. In addition, the rubber valve on the end of the evaporator drain must be in place to prevent water from being carried to the overhead vents. A clogged drain can normally be opened up with a slender piece of wire or with low-pressure shop air. The evaporator drain should be checked during routine maintenance inspections. Washing the evaporator core with a water solution and a commercially available cleaner can rectify a mildew smell that is noticeable during A/C system operation. Some manufacturers make an aerosol fungicide that can be sprayed onto the evaporator core surface. Access to the evaporator core can be gained by opening the return air vent cover.

Task B3.9 Identify and inspect A/C system service valves and connections; repair as required.

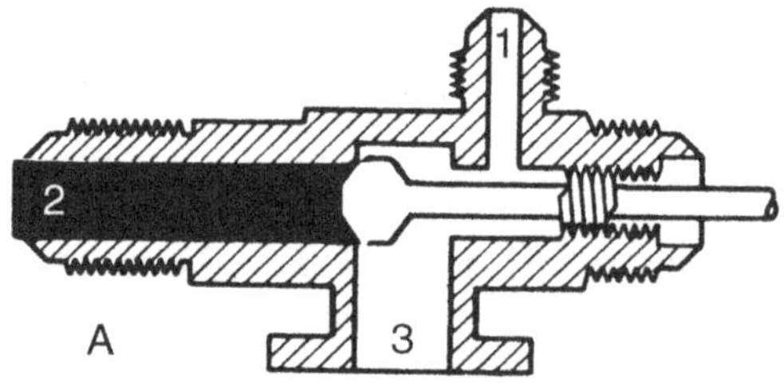

Front-Seated Position

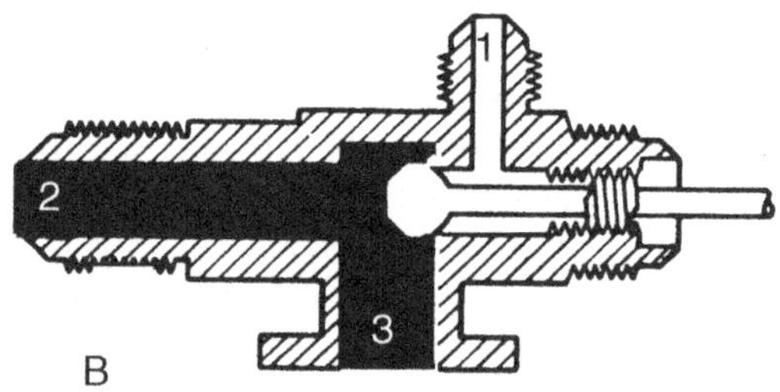

Back-Seated Position

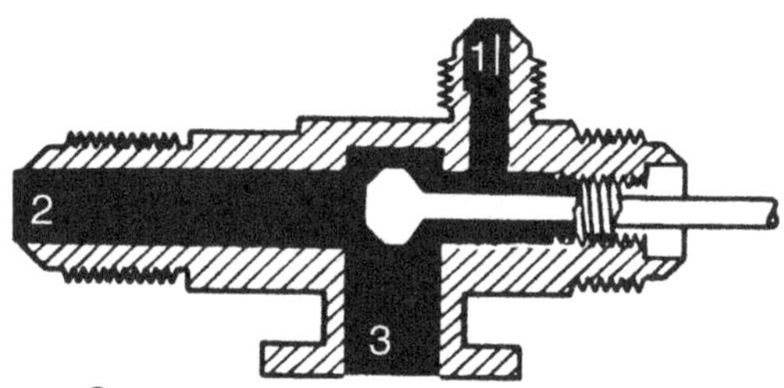

Mid-Positioned (Cracked)

A/C system service valves are used in bus A/C systems. In the figure, diagram B shows the normal operation, or back-seated position, where the valve stem is rotated counterclockwise to seat the rear valve face and seal off the service gauge port. Diagram C shows the mid-position used during A/C system diagnosis and service. Diagram A shows the front-seated position (valve stem rotated clockwise to seat the front valve face) that isolates the compressor from the A/C system. This position allows a technician to service the compressor without discharging the entire system. The system must never be operated with the discharge service valve in the front-seated position.

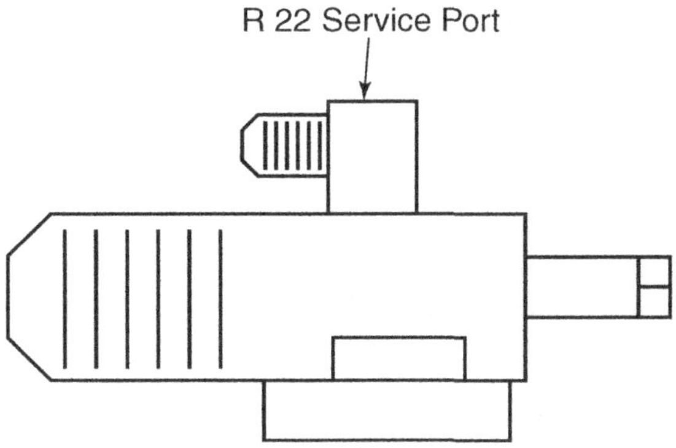

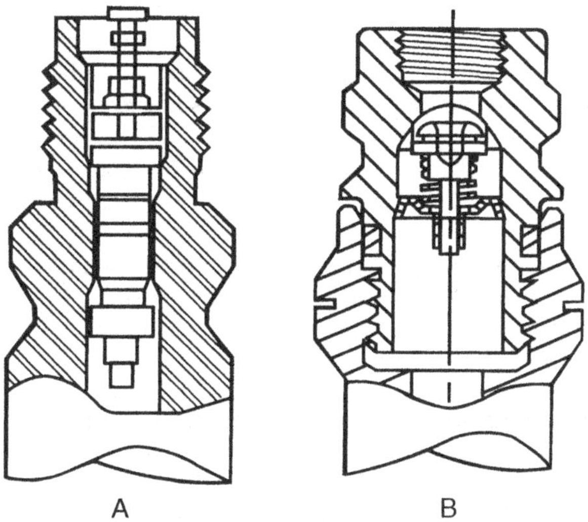

Most R-22 A/C systems have service ports similar to diagram A in the figure above, which feature external threads and replaceable Schrader valves like the ones used in tire valve stems. R-134a service ports have internal threads and are a quick-disconnect design, like the one shown in diagram B. The low-side port is 13 mm in diameter and the high-side port is 16 mm in diameter.

Task B3.10 Diagnose cause of A/C system failures which result in refrigerant loss from the high-pressure relief device.

All mobile A/C systems are equipped with a high-pressure relief device. In most systems, this device is a self-resetting relief valve, which is threaded into the high side of the compressor. The high-pressure relief valve will vent refrigerant from the system in the event that high-side pressure exceeds safe levels. This is often an overlooked area when total refrigerant loss has occurred.

C. Heating and Engine Cooling Systems Diagnosis, Service, and Repair (8 Questions)

Task C1 Diagnose the cause of outlet air temperature control problems in the HVAC system; determine needed repairs.

Heating system design achieves heater temperature control by air modulation and coolant flow control. In a bus system, coolant flows through the heater core at a constant rate. Air flows through the HVAC through the heater core. In a coolant flow control system, temperature control is achieved by using a coolant control valve (hot-water valve) to limit the amount of coolant allowed to flow through the heater core, depending on the temperature setting. The coolant control valve is electrically operated.

Task C2 Diagnose window fogging problems; determine needed repairs.

Windshield fogging may be caused by a leaking heater core or by a clogged evaporator case drain. If windshield fogging is accompanied by the smell of antifreeze, the heater core is at fault. If you smell a pungent odor, then the evaporator case drain may be clogged.

Task C3 Perform engine cooling system tests for leaks, protection level, contamination, coolant level, temperature, and conditioner concentration; determine needed repairs.

To test the engine cooling system for leaks, remove the radiator pressure cap following the manufacturer's instructions. Replace the cap with a standard cooling system pressure tester and pressurize the system to operating pressure. Perform a careful visual inspection for leaks. Often you can confirm cooling system contamination with a visual inspection. Rust in the system will turn the coolant an opaque reddish-brown color. If engine oil or transmission fluid has entered the system, the coolant will contain thick deposits resembling a milk shake. Coolant freeze protection level is most easily determined using a cooling system hydrometer or refractometer, which is preferred by some engine manufacturers.

To check the concentration of the cooling system conditioner, use test strips that match the type of conditioner being used and then add conditioner as needed. Be careful not to over-condition the coolant because it can lead to conditioner dropout and system damage. Cooling system temperatures are easily checked using a gauge known to be accurate or a temperature probe.

Task C4 Inspect engine cooling and heating system hoses, lines, and clamps; replace as required.

Cooling system and heater hoses must be replaced if they are cracked or brittle, or if they show signs of bulging or abrasion. Hoses begin to deteriorate from the inside first, so examine the inside of any suspected hoses to determine their true condition. Often a cooling system leaks only when cold, which is known as a cold-water leak. The use of constant torque clamps will minimize this. Hose clamps should be replaced if they are deformed or cracked, or if they cannot be operated smoothly.

Task C5 Inspect and test radiator, pressure relief devices, and coolant recovery system (surge tank); determine needed repairs.

The technician should inspect the radiator at every scheduled preventive maintenance inspection. Check the radiator for bent fins, kinked or cracked tubes, and leaks. With the engine at normal operating temperature, the temperature of the radiator core should be uniform. Cool spots indicate clogged tubes in the core. You test the pressure relief valve for pressure using a cooling system pressure tester. If the pressure cap seal, sealing gasket, or seat is damaged, the engine will overheat and coolant will be blown out of the surge tank. Replace the cap gasket if it does not hold the pressure specified by the manufacturer or if there is any sign of physical damage. The coolant surge tank should be checked for leaks and for sediment buildup, and should be repaired or cleaned as necessary.

Task C6 Inspect water pump and drive system; determine needed repairs.

The coolant circulating pump is driven by an electric motor and used to move coolant through the HVAC system. The most apparent sign of water pump failure is fluid loss from the weep hole area. A visual inspection for leaks from the hoses and pump housings should be performed at each inspection interval.

Task C7 Inspect and test thermostats, by-passes, housings, and seals; determine needed repairs.

The only accurate way to test a thermostat is to remove it from the vehicle, place it in a container of water with a thermometer, and heat the container until the thermostat opens. The water temperature at the point when the thermostat opens should equal the manufacturer's specification for thermostat opening temperature. Other indications that the thermostat has opened include visible coolant flow in the upper radiator tank, a hot upper radiator hose, and an engine temperature gauge that indicates normal operating temperature.

Task C8 Flush and refill cooling system; bleed air from system.

The cooling system should be flushed if there is any sign of rust or contamination in the coolant. After flushing, the entire system should be drained and fresh coolant should be added. The technician should consult the service manual to verify cooling system capacity and bleeding procedure.

Task C9 Inspect and test fan, fan drives, controls, and fan shroud; repair or replace as required.

The technician should inspect the cooling fan operation. It should be repaired if the fan speed is not correct for the temperature. The bus fan system is hydraulically operated and has a safety system that causes the fan to stop when a fire is detected.

Task C10 Inspect and test heating system coolant control valve(s) and manual shut-off valves; replace as required.

The coolant control valves (hot-water valve) controls the flow of coolant through the heater core and driver heater. The coolant control valve may be electrically operated or cable operated. One can verify proper valve operation by manually opening and closing the valve and observing the temperature change in the downstream heater hose.

Task C11 Inspect and flush driver's heater and/or defroster cores; replace as required.

In a poorly maintained cooling system, sediment may build up in the heater core, causing poor heater performance. Flushing the heater core will restore heater efficiency and may reveal small heater core leaks. The heater core can be pressure tested independently from the rest of the cooling system; however, the core should never be pressurized in excess of normal operating pressure. When replacing a heater core, it is important to reinstall all foam mounting insulators if used to minimize the risk of vibration damage and to ensure a good seal around the heater core.

Task C12 Perform heating system tests to include coolant flow and booster pump(s) operation.

The booster pump is critical to the operation of the heating system. The engine water pump will not provide enough power to circulate the coolant through the entire bus. The boost is usually an electrically powered DC or an ECDC unit with a direct or magnetically coupled pump.

Task C13 Inspect, test, and diagnose heater/defroster and defogger system problems; check blowers, fans, resistors, switches, relay/modules, sensors, wiring, and protection devices; repair or replace as required.

Most heater defroster systems have two blower motors, one for the left side and one for the right side. The motors usually have two or three speeds. Resistors are sometimes used to control voltage to the motors. Cable-controlled dampers direct the airflow to the defroster and driver heat vents.

Task C14 Inspect, test, and diagnose service heating, ventilating, and defrosting control panel assemblies, cables, and linkages; repair or replace as required.

The HVAC display and control module is often separate from the driver heater/defroster controls. The heater/defroster is usually controlled by electrical switches for the fans and knobs that are pulled to desired heat or defrost positions or by knobs that are turned to control heat or defroster airflow. Usually the dampers are cable actuated.

Task C15 Inspect, test, and diagnose heating, ventilating, and defrosting control switches, hoses, and solenoid valves; repair or replace as required.

The heater, ventilator and defroster may have an electric coolant valve operated by electrical rotary or toggle switches. The heat is supplied via a heater core mounted in the driver heater assembly. The coolant hoses and connections need to be checked periodically for leaks and deterioration. The electrical operation of fans needs to be tested periodically to assure that all motors and all speeds are operating.

Task C16 Inspect, test, and adjust heating, ventilating, and defrosting ducts, filters, doors, cables, linkages, hoses, and outlets; repair or replace as required.

The HVAC system has two major components. The driver heater/defroster and the main HVAC unit —either the rear insert or roof-top unit. The control switches consist of electrical switches and cable-controlled doors; alternatively, the damper or doors may have electric actuators.

The cable controls are attached to valves and door levers. The levers may control microswitches that control electric solenoid valves for coolant or booster pumps. The cables sometimes break or become loose or inoperative. The periodic inspection of the HVAC system should include lubricating the cables and operating the doors for air distribution. The temperature of the coolant tubing should indicate the operation of the coolant valve, whether electric solenoid or lever-controlled.

Task C17 Test, diagnose, and service fuel-fired auxiliary heater system.

The use of an auxiliary heater in transit buses is necessary in northern climates because the engine cannot produce enough heated water for the HVAC system. The fuel-fired heaters are usually about 8,000 BTU capacity. They are thermostatically controlled and either internally controlled by microprocessors or electronically controlled.

D. Heating, Ventilation, and Air Conditioning Operating Systems and Related Controls Diagnosis and Repair (11 Questions)

1. Electrical Systems (7 Questions)

Task D1.1 Diagnose the cause of failures in HVAC electrical control systems; determine needed repairs.

Diagnosing HVAC electrical control system problems is no different than diagnosing other electrical systems. The technician should follow a systematic approach to troubleshooting, including verifying the concern and performing a thorough visual inspection. The technician should use all available resources, including electrical schematic diagrams and service manual diagnostic routines, to locate and repair the cause of the concern.

Task D1.2 Inspect and test HVAC defroster blower motors, resistors, switches, relays, modules, wiring, and protection devices; repair or replace as required.

A 65 amp circuit breaker protects the HVAC blower circuit. Many blower systems are powered by one or more relays. The operator selects the desired blower motor speed by using the blower switch. The blower resistor block or internal wiring in the motor control the speed. The resistors are used to step-down the voltage to the blower motor, thereby providing multiple blower speeds.

Task D1.3 Inspect and test HVAC compressor clutch relays, modules, wiring, sensors, switches, diodes, and protection devices; repair or replace as required.

The A/C compressor clutch coil is usually energized by an electronically controlled relay. Some blower motor circuits connect the motor switch and resistor branches to the positive side of the motor. When the system is on, voltage is applied to the resistor then to the motor assembly. When you set the blower to high, current passes through the high-speed contacts in the high-speed relay. Some motors use internal windings to control the speed of the blowers. The latest technology uses A/C current from motor drivers/inverters and controls the speed by varying the frequency of the A/C current.

Task D1.4 Inspect and test HVAC-related electronic engine control systems; determine needed repairs.

A/C compressor clutch operation may be dependent upon signals from various system sensors. If the compressor discharge pressure or the temperature is too high, the clutch will be disengaged.

Task D1.5 Inspect and test HVAC evaporator/heater and condenser fan motors, motor drivers (alternating current and direct current) relays, modules, switches, sensors, wiring, and protection devices; repair or replace as required.

Buses are equipped with electric condenser fans. Electric fans are usually controlled by an electronic relay, which is in turn controlled by the HVAC control module (ECM). The ECM grounds the low- and high-speed fan relays in response to engine coolant temperature and compressor head temperature.

Compressor head temperature switches, when used, are mounted so that they contact the compressor case. When they sense that the temperature is exceeding a certain threshold, they will open, turning the compressor clutch off.

Task D1.6 Inspect and test HVAC system electrical control panel assemblies; repair or replace as required.

Electrical control panel assemblies for manual and some electronically controlled HVAC systems are modular in design, allowing for replacement of individual relays without replacing the entire panel. Most electronic control panels allow the technician access only to replace relays.

2. Constant/Automatic Temperature Control Systems (4 Questions)

Task D2.1 Diagnose constant/automatic temperature control system problems; determine needed repairs.

Most microprocessor systems provide internal diagnostic capabilities. On some systems, diagnostic trouble codes may be displayed digitally on the control panel; on other systems, a hand-held scan tool or PC interface must be used to retrieve codes. The technician should always refer to diagnostic routines in the correct service manual when attempting to troubleshoot trouble codes. Most importantly, the technician must rule out the possibility of mechanical failures before searching for electronic malfunctions.

Task D2.2 Inspect and test climate control temperature sensors; repair or replace as required.

ATC systems rely on a variety of sensors to provide feedback to the ATC control unit. The control unit uses the sensor signals to determine how much heating or cooling is required to maintain the desired interior temperature. The ambient temperature sensor monitors outside air temperature. The engine coolant temperature sensor monitors engine coolant temperature. The ATC temperature (or return air) sensor monitors the temperature of the air in the interior. This automatically controls the temperature in the bus.

Task D2.3 Inspect, test, and adjust heater coolant valve and controls; repair or replace as required.

Transit bus HVAC systems use a water valve to control water flow through the heater core. They are prone to a number of problems. The valves may become clogged with deposits and seize. They should also be checked for flow restrictions. Replace the valves if they do not perform. Cable-operated valves can suffer from bent, kinked, or misadjusted cables. Air pressure-operated water valves may leak air pressure into the coolant.

Task D2.4 Inspect and test solenoids and switches; repair or replace as required.

In most ATC systems, the mode doors are controlled using electronic actuators that are similar to the blend door actuator. Faulty mode door actuators are usually diagnosed by following diagnostic routines in the vehicle service manual.

Task D2.5 Inspect and test constant/automatic temperature control panels; repair or replace as required.

The only serviceable components in most ATC control panels are illumination bulbs. Faulty ATC control panels are generally diagnosed by following diagnostic routines in the vehicle service manual, or with an electronic service tool.

Task D2.6 Inspect and test constant/automatic temperature control microprocessor (climate control computer/programmer); repair or replace as required.

The ATC control unit is a stand-alone component. The control unit is generally multiplexed to the bus system control units. Electronic diagnostic capabilities record failure codes in case of control unit failure.

Task D2.7 Connect data reader/computer to determine fault codes and perform system tests; check and adjust system parameters; clear fault codes.

Current HVAC systems are microprocessor controlled and have the ability to provide diagnostic codes to enable troubleshooting the systems and to change the operating parameters. Some systems use a laptop computer to access the unit processor and make program or parameter changes and to clear codes. Other systems offer the diagnostics displayed on the control display. Each system has the procedures within the service manuals to either plug in the laptop or access the diagnostics through the operator display panel.

E. Refrigerant Recovery, Recycling, and Handling (5 Questions)

NOTE: Tasks 1 through 5 should be accomplished in accordance with published EPA and appropriate SAE "J" standards for R-22, R-134a, R-407c, and EPA-approved refrigerant blends. Service must be performed by EPA certified technicians.

Task E1 Maintain and verify correct operation of certified equipment.

The Clean Air Act (CAA) establishes the following rules for record keeping and operation of certified refrigeration service equipment:

1. Any person who owns approved refrigerant recycling equipment certified under the Act must maintain records of the name and address of any facility to which refrigerant is sent.
2. Any person who owns approved refrigerant recycling equipment must retain records demonstrating that all persons authorized to operate the equipment are certified under the Act.
3. Public Notification—Any person who conducts any retail sales of a Class I or Class II substance must prominently display a sign that reads: "It is a violation of federal law to sell containers of Class I and Class II refrigerant of less than 20 pounds of such refrigerant to anyone who is not properly trained and certified."
4. Any person who sells or distributes any Class I or Class II substance that is in a container of less than 20 pounds of such refrigerant must verify that the purchaser is certified, and must retain records for a period of three years. (These records must be maintained on-site.)

Task E2 Identify and recover A/C system refrigerant.

According to Department of Transportation (DOT)/Air Conditioning and Refrigeration Institute (ARI) guidelines, 4B4 cylinders used to store recovered refrigerant shall be painted gray with the top shoulder portion painted yellow. The refrigerant type to be stored in a given container must be clearly marked on the container's label. For recovery/recycling purposes, only cylinders that are identified for recovered refrigerant may be used. Never use a cylinder that is intended to contain new refrigerant to store recovered refrigerant. Returnable/reusable cylinders meet DOT specification 4BA-300. These cylinders are characterized by a combined liquid/vapor valve located on top.

Task E3 Recycle refrigerant.

Differences between the terms recover, recycle, and reclaim must be completely understood and properly used within the industry. To recover refrigerant is to remove refrigerant in any condition from a system and store it in an external container. The refrigerant must then either be recycled on-site or shipped off-site for reclamation. To recycle refrigerant is to reduce contaminants in used refrigerant by oil separation with single or multiple passes through devices such as replaceable filter driers, which reduce moisture, acidity, and particulate matter. To reclaim refrigerant is to reprocess refrigerant to new product specifications by various means, which may include distillation. Chemical analysis of the refrigerant is required to assure that appropriate product specifications are met. Reclamation usually implies the use of procedures available only at processing or manufacturing facilities.

Task E4 Handle, label, and store refrigerant.

Any portable container used for transfer of reclaimed or recycled refrigerant must conform to DOT and Underwriters Laboratories (UL) standards. Before introducing refrigerant into an approved storage cylinder, the cylinder must be evacuated to at least 27 inches of mercury. Cylinder-safe filling level must be monitored by measured weight. Shut-off valves are required within 12 inches (30 cm) of service hose ends. Shut-off valves must remain closed while connecting and disconnecting hoses to vehicle air conditioning service ports. Safety goggles should always be worn while working with or around refrigerant.

Task E5 Test recycled refrigerant for non-condensable gases.

To test a refrigerant for non-condensable gases, compare the pressure of the refrigerant in a cylinder to the theoretical pressure of pure refrigerant at a given temperature. If the actual pressure in the cylinder is higher than the theoretical pressure, the refrigerant is contaminated with non-condensable gas.

Pressure-Temperature Relationship

Temperature ° F (°C)	R-12 PSIG	R-12 (bar/kg/cm2)	R-134 A PSIG	R-134 A (bar/kg/cm2)
–15 (–26.1)	2.5	(.17/.18)	0	(0)
–10 (–23.3)	4.5	(.31/.32)	2.0	(.14/.14)
–5 (–20.5)	6.7	(.46/.03)	4.1	(.28/.29)
0 (–17.8)	9.2	(.63/.65)	6.5	(.45/.46)
5 (–15.0)	11.8	(.81/.83)	9.1	(.63/.64)
10 (–12.2)	14.7	(1.0/1.0)	12.0	(.89/.84)
15 (–9.4)	17.7	(1.2/1.2)	15.1	(1.0/1.2)
20 (–6.7)	21.1	(1.5/1.5)	18.4	(1.3/1.3)
25 (–3.9)	24.6	(1.7/1.7)	22.1	(1.5/1.6)
30 (–1.1)	28.5	(2.0/2.0)	26.1	(1.8/1.8)
35 (1.7)	32.6	(2.2/2.3)	30.4	(2.1/2.1)
40 (4.4)	37.0	(2.6/2.6)	35.0	(2.4/2.5)
45 (7.2)	41.7	(2.9/3.0)	40.0	(2.8/2.8)
50 (10.0)	46.7	(3.2/3.3)	45.4	(3.1/3.2)
55 (12.8)	52.1	(3.6/3.7)	51.2	(3.5/3.6)
60 (15.6)	57.8	(4.0/4.1)	57.4	(4.0/4.0)
65 (18.3)	63.8	(4.4/4.5)	64.0	(4.4/4.5)
70 (21.1)	70.2	(4.8/5.0)	71.1	(5.0/5.0)
75 (23.9)	77.0	(5.3/5.4)	78.6	(5.4/5.5)
80 (26.7)	84.2	(5.8/6.0)	86.7	(6.0/6.1)
85 (29.4)	91.7	(6.3/6.4)	95.2	(6.6/6.7)
90 (32.2)	99.7	(6.9/7.0)	104.3	(7.2/7.3)
95 (35.0)	108.2	(7.5/7.6)	113.9	(7.9/8.0)
100 (37.8)	117.0	(8.1/8.2)	124.1	(8.6/8.7)
105 (40.6)	126.4	(8.7/8.9)	134.9	(9.3/9.5)
110 (43.3)	136.2	(9.4/9.6)	146.3	(10.1/10.3)
115 (46.1)	146.5	(10.1/10.3)	158.4	(11.0/11.1)
120 (48.9)	157.3	(11.0/11.1)	171.1	(11.8/12.0)

Task E6 Follow federal and local laws for service procedures.

When retrofitting systems to use R-134a refrigerant, it is recommended that procedures be followed as outlined by OEMs and be in compliance with all federal and local statutes. This includes recovery, evacuation, and replacement of components.

5 Sample Test for Practice

Sample Test

Please note the letter and number in parentheses following each question. They match the task in Section 4 that discusses the relevant subject matter. You may want to refer to the overview using the cross-referencing key to help with questions posing problems for you.

1. All of the following statements about charging an A/C system are true EXCEPT:
   A. refrigerant may be installed through both service valves when the engine is not running.
   B. refrigerant may be installed through the low-side service valve when the engine is running.
   C. refrigerant may be installed through both service ports when the engine is running.
   D. you may install refrigerant directly from an approved charging station. (B1.8)

2. A growling noise is coming from the water pump. Technician A says that the water pump bearing could be the cause. Technician B says that the impeller has been eroded by cavitation. Who is correct?
   A. A only
   B. B only
   C. Both A and B
   D. Neither A nor B (C6)

3. During normal A/C operation, a loud hissing noise is heard and a cloud of vapor is discharged from the HVAC compartment. Technician A says that excessive high-side pressure that caused the pressure relief valve on the A/C compressor to trip may have been the result of over pressure. Technician B says that the relief valve may have tripped due to a defective high-pressure valve. Who is correct?
   A. A only
   B. B only
   C. Both A and B
   D. Neither A nor B (B2.1)

4. A heater does not supply the bus with enough heat. The coolant level and blower test OK. Technician A says an improperly adjusted temperature could be the cause. Technician B says a clogged heater core could be the cause. Who is correct?
   A. A only
   B. B only
   C. Both A and B
   D. Neither A nor B (C1)

5. Before a portable container is used to transfer recycled R-12, it must be evacuated to at least:
   A. 20 in. Hg.
   B. 22 in. Hg.
   C. 27 in. Hg.
   D. 12 in. Hg. (E4)

6. Which is the best method of removing particulate from an A/C system after a mechanical failure?
   A. A/C flush solvent in the reverse flow
   B. R-11 flushing
   C. Inline suction filter
   D. Nitrogen flushing (B1.7)

7. Technician A says that deformed or improperly aligned condenser mounting will damage the A/C compressor. Technician B says that deformed or improperly aligned condenser mounting could damage the condenser and refrigerant lines. Who is correct?
   A. A only
   B. B only
   C. Both A and B
   D. Neither A nor B (B3.4)

8. A sweet odor comes from the overhead vents while operating the A/C in cool mode. Technician A says he can smell R-134a in the bus, so the A/C system must have a leaking evaporator. Technician B says the heater core could be leaking. Who is correct?
   A. A only
   B. B only
   C. Both A and B
   D. Neither A nor B (A2)

9. Technician A says that whenever the A/C system has been open to the atmosphere, the dryer should be replaced. Technician B says that when a compressor is replaced, the dryer should also be replaced. Who is correct?
   A. A only
   B. B only
   C. Both A and B
   D. Neither A nor B (B3.5)

10. Technician A says that when installing a new or rebuilt compressor, you should first turn it over by hand and drain any oil shipped with the compressor, then add the correct amount of the specified oil before installing it. Technician B says that some compressors are shipped without oil. Who is correct?
    A. A only
    B. B only
    C. Both A and B
    D. Neither A nor B (B2.6)

11. The process that reduces contaminants in used refrigerant by using oil separation and filter core dryers is:
    A. restoration.
    B. recovery.
    C. recycling.
    D. reclamation. (E3)

12. While testing an A/C system at normal speed and temperature, the high- and low-side pressure are the same. Technician A says the clutch may not be engaged. Technician B says there may be a restriction in the expansion valve. Who is correct?
    A. A only
    B. B only
    C. Both A and B
    D. Neither A nor B (B1.3)

13. Technician A says that the heater control valve can be operated by engine speed. Technician B says that the heater control valve can be cable operated. Who is correct?
    A. A only
    B. B only
    C. Both A and B
    D. Neither A nor B (C10)

14. A storage container or air conditioning system containing R-134a (at rest) and subject to an ambient temperature of 70° F will have an internal gauge pressure of approximately (see table):
    A. 220 psi.
    B. 125 psi.
    C. 70 psi.
    D. 30 psi. (E5)

15. All of the following can cause A/C compressor drive belt misalignment EXCEPT:
    A. a bent compressor mounting bracket holder.
    B. a faulty compressor clutch bearing assembly.
    C. an improperly set compressor clutch air gap.
    D. a faulty idler pulley. (B2.3)

16. To check and adjust the A/C compressor lubricant level:
    A. quickly purge the system and add oil charges to refill it.
    B. open the drain plug and crank the engine until the compressor is empty, then pump fresh oil into the compressor.
    C. remove the compressor from the vehicle, drain the oil, and add the specified quantity of fresh oil.
    D. add refrigerant oil until you can see the oil level with system running and stabilized. (B2.5)

17. A cracked A/C compressor mounting plate could cause all of the following symptoms EXCEPT:
    A. drive belt wear.
    B. internal compressor damage.
    C. vibration with the A/C compressor clutch engaged.
    D. drive belt squeal or chatter. (B2.7)

18. Technician A says the R-22 A/C service valve has a removable core. Technician B says the R-134a service valve must be rear-seated during A/C compressor replacement. Who is correct?
    A. A only
    B. B only
    C. Both A and B
    D. Neither A nor B (B3.9)

19. The refrigerant in an A/C system starts changing state from liquid to vapor as it flows through the:
    A. vapor line.
    B. condenser.
    C. expansion valve.
    D. capillary tube. (B3.7)

20. A band of frost on the A/C high-pressure hose, upstream from the orifice tube, indicates:
    A. a defective compressor discharge valve.
    B. a restriction in the high-pressure hose.
    C. a clogged expansion valve.
    D. moisture in the system. (B1.4)

21. While using an electronic leak detector, where should the detector's probe be placed near a suspected leak in order to most effectively locate the leak?
    A. Directly above the suspected leak area
    B. Directly below the suspected leak area
    C. Six inches upstream from the suspected leak area
    D. Six inches downstream from the suspected leak area (B1.5)

22. The circuit breaker that controls power to the defroster motor switch is tripped. Technician A says the defroster switch may be bad. Technician B says the wiring to the switch or motor could be bad. Who is correct?
    A. A only
    B. B only
    C. Both A and B
    D. Neither A nor B (C15)

23. A driver complains of low airflow to the windshield. A quick examination proves the driver correct. Technician A discovers that the heater filter is very dirty and says this is the problem. Technician B says the filter can't cut airflow that much. Who is correct?
    A. A only
    B. B only
    C. Both A and B
    D. Neither A nor B (C16)

24. A typical microprocessor-controlled system will delay blower motor operation until coolant temperature reaches:
    A. 90° F (32° C).
    B. 70° F (21° C).
    C. 105° F (40° C).
    D. 120° F (49° C). (D2.2)

25. The best instrument to use when troubleshooting an A/C electronic circuit with solid-state components is:
    A. a digital multimeter (DMM).
    B. a self-powered test lamp.
    C. an analog volt/ohmmeter.
    D. a 12 V test lamp. (D2.6)

26. A driver report indicates poor heat to the driver heater. The technician discovers the breaker for the booster pump is tripped. Technician A says the breaker should be reset and the bus returned to service. Technician B says the booster pump should be checked for high current draw. Who is correct?
    A. A only
    B. B only
    C. Both A and B
    D. Neither A nor B (C12)

27. During a bus inspection a technician discovers that only the left blower motor is operating. The right motor does not operate at high or low speeds. The correct course of action should be to:
    A. check the current to the motor.
    B. check the blower motor switch.
    C. change the blower motor resistor.
    D. change the blower motor. (C13)

28. Owners of approved refrigerant recycling equipment must maintain records that demonstrate:
    A. only certified technicians operate the equipment.
    B. the equipment is operated under the supervision of certified technicians.
    C. technicians who are undergoing certification training operate the equipment.
    D. the equipment is operated only when a certified technician is on the premises. (E1)

29. The right-side windshield is frosted up. An inspection determined that the defroster damper is not working. Technician A says that the cable is loose or broken at the lever connection. Technician B says that the water valve to allow hot coolant to the heater/defroster core is not working. Who is correct?
    A. A only
    B. B only
    C. Both A and B
    D. Neither A nor B (C14)

30. The A/C high-pressure relief valve shows evidence of slight oil leakage. Technician A says you must replace the valve and repair the leak. Technician B says that if the valve is just leaking a little oil and not refrigerant, repair is not necessary. Who is correct?
    A. A only
    B. B only
    C. Both A and B
    D. Neither A nor B (B3.10)

31. The lubricant used in R-134a Type 2 A/C systems is:
    A. a polyolester (POE)-based lubricant.
    B. a mineral-based petroleum lubricant.
    C. DEXRON or DEXRON II lubricant.
    D. type C-3, SAE 30-based oil lubricant. (B1.9)

32. While a unit is operating in the cool mode, the system shuts down and logs a code 10. Technician A says the unit has a high discharge code and the unit should be shut off and the code cleared and put back into service. Technician B says the high discharge could be from a blocked condenser or a bad condenser fan. Who is correct?
    A. A only
    B. B only
    C. Both A and B
    D. Neither A nor B (A4)

33. Technician A says that a fan operating at maximum rpm all the time has a defective fan control. Technician B says that the fan is operating correctly. Who is correct?
    A. A only
    B. B only
    C. Both A and B
    D. Neither A nor B (C9)

34. While testing a heater core for leaks, you apply 10 psi of air pressure. The pressure bleeds to 5 psi in three minutes. You should conclude that:
    A. the core is OK.
    B. the core may require repair in future.
    C. the core leaks and needs repair now.
    D. this is not a valid test for a leaking heater core. (C11)

35. All of the following are good methods of verifying that an engine thermostat opens EXCEPT:
    A. feeling the upper radiator hose.
    B. watching the temperature gauge.
    C. watching for motion in the upper radiator tank.
    D. watching the computer read-out of the engine temp sensor. (C7)

36. A driver complains that he is unable to change the output temperature of the HVAC system. The Most-Likely cause of this problem is:
    A. a stuck coolant valve.
    B. a defective compressor clutch.
    C. a clogged expansion valve screen.
    D. a defective blower switch. (B1.1)

37. With the driver heater system in DEFROST mode, the blower on HIGH, and the temperature control on COLD, the bottom of the windshield fogs up on the outside because:
    A. the evaporator case drain is clogged.
    B. the heater core leaks.
    C. the evaporator core is iced up.
    D. the cold windshield causes moisture to condense from the outside air. (C2)

38. What evidence should a technician look for to determine if a bus has had a retrofit from R-12 to R-134a?
    A. The compressor has been replaced.
    B. The dryer has been replaced.
    C. The caps are missing from the service ports.
    D. A retrofit decal is installed on the receiver tank. (E6)

39. A driver report motes that the HVAC system has failed in the cool mode and the warning indicator is red. Technician A is troubleshooting the system and discovers an alarm code 10 is present. Technician A indicates that the system failed because of a condenser fan failure. Technician B says the system failed because of an A/C motor driver failure. Who is correct?
    A. A only
    B. B only
    C. Both A and B
    D. Neither A nor B (D2.7)

40. Which of the following should be used to lubricate replacement A/C hose O-rings?
    A. Petroleum jelly
    B. Transmission fluid
    C. Silicone grease
    D. Refrigeration oil (B3.2)

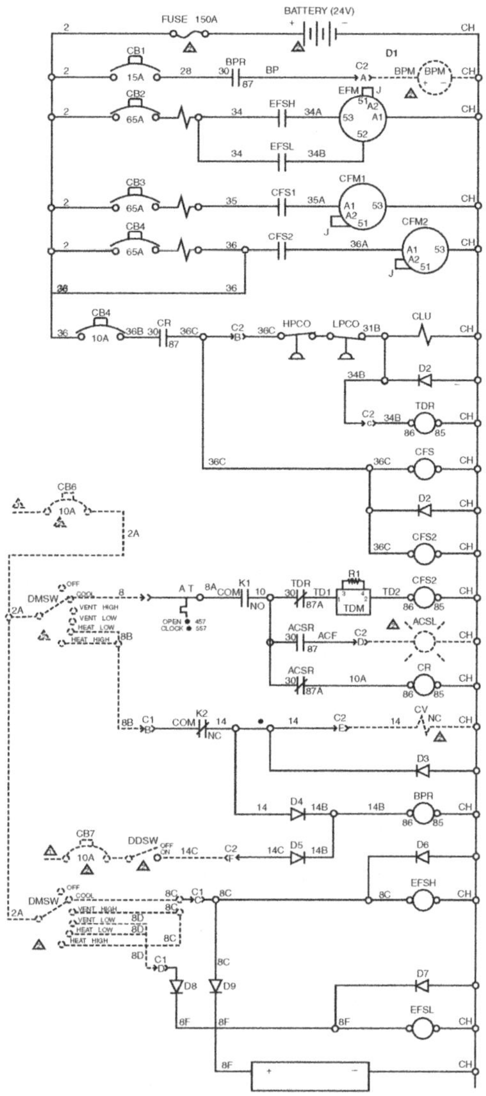

LEGEND		LOCATION
ACSR	AC STOP LIGHT RELAY	16, 17, (15)
ACSL	AC STOP LIGHT	16
AT	AMBIENT THERMOSTAT SWITCH	15
BATT	24VDC BATTERY	2
BPM	BOOST PUMP MOTOR	3
BPR	BOOST PUMP RELAY	3, (20)
C1	CONNECTOR 1	15, 18, 22, 24, 27, 28
C2	CONNECTOR 2	3, 9, 11, 16, 18, 21
CB-1-7	CIRCUIT BREAKERS	3, 4, 5, 6, 7, 9, 13, 21
CFM1	CONDENSER FAN MOTOR 1	6
CFM2	CONDENSER FAN MOTOR 2	7
CFS1	CONDENSER FAN SOLENOID 1	6, (12)
CFS2	CONDENSER FAN SOLENOID 2	7, (14)
CLU	COMPRESSOR CLUTCH	9
CR	CLUTCH RELAY	9, (17)
CV	ENGINE COOLANT VALVE (NC)	18
D1–D4	DIODES	10. 13, 19, 20
D5–D9	DIODES	21, 21, 24, 24
DMSW	DRIVERS MODE SWITCH	15, 22
SPC	SET POINT CONTROL	27
ECT	ELECTRIC CONTROL THERMOSTAT	27
EFM	EVAPORATOR FAN MOTOR	4
EFSH	EVAPORATOR FAN SOLENOID HIGH	4, (22)
EFSL	EVAPORATOR FAN SOLENOID LOW	5, (25)
FUSE	FUSE 15# AMP	2
HPCO	HIGH PRESSURE CUTOUT SWITCH	9
K1	THERMOSTAT COOL RELAY	15, (29)
K2	THERMOSTAT HEAT RELAY	18, (29)
LPCO	LOW PRESSURE CUTOUT SWITCH	9
R1	RESISTOR <100K HOMS>	15
TDM	TIME DELAY MODE	15
TDR	TIME DELAY RELAY	15, (11)
TS	THERMOSTAT SENSOR	27

41. Refer to the figure. When the blower switch is in the high position, which relay contacts are used to control blower motor speed?
A. EFSH
B. EFSL
C. CFS1
D. None (D1.2)

42. Technician A says the HVAC control panel sometimes contains replaceable components such as relays. Technician B says that you must replace the HVAC control panel as a unit if any of the components fail. Who is correct?
A. A only
B. B only
C. Both A and B
D. Neither A nor B (D1.6)

43. All of these statements about cooling system service are true EXCEPT:
A. when the cooling system pressure is increased, the boiling point is decreased.
B. if more antifreeze is added to the coolant mix, the boiling point is increased.
C. a good quality ethylene glycol antifreeze contains a corrosion inhibitor.
D. coolant solutions must be recovered, recycled, or handled as hazardous material. (C8)

44. Raising the pressure in the cooling system:
    A. lowers the boiling point of the coolant.
    B. raises the boiling point of the coolant.
    C. prevents corrosion in the cooling system.
    D. does not affect the boiling point of the coolant. (C5)

45. A driver reports excessive cycling of the compressor. The bus is reaching set point and is comfortable. Technician A says the unloader is not working correctly. Technician B says the bus is low on refrigerant. Who is correct?
    A. A only
    B. B only
    C. Both A and B
    D. Neither A nor B (B2.8)

46. The recommended tool for diagnosing electronic sensors and actuators in an ATC system is:
    A. a 12-volt test light.
    B. an analog multimeter.
    C. a digital multimeter (DMM).
    D. an A/C system charging station. (D2.4)

47. Technician A says that some HVAC control panels can be recalibrated using a hand-held digital diagnostic tool. Technician B says that a hand-held digital diagnostic tool can be used to help diagnose HVAC system failures. Who is correct?
    A. A only
    B. B only
    C. Both A and B
    D. Neither A nor B (D2.1)

48. An HVAC system outputs air at a constant temperature, regardless of the temperature setting. The Most-Likely cause is:
    A. a low refrigerant charge.
    B. low coolant level.
    C. a compressor clutch failure.
    D. a stuck coolant valve. (A3)

49. The supply heater hose has a slight bulge. Technician A says that the hose does not need to be replaced unless it appears cracked. Technician B says that the bulge indicates a weak spot and the hose should be replaced. Who is correct?
    A. A only
    B. B only
    C. Both A and B
    D. Neither A nor B (C4)

50. The low-pressure cutout switch senses pressure in the:
    A. system high side.
    B. atmosphere.
    C. system low side.
    D. bus interior. (B2.2)

51. One segment of the digital readout on an ATC control panel is inoperative. Technician A says that the control panel should be replaced. Technician B says that the Light Emitting Diode (LED) can be replaced. Who is correct?
    A. A only
    B. B only
    C. Both A and B
    D. Neither A nor B (D2.5)

52. Refrigerant oil must be added to the system after a new a component when replacing which of the following?
    A. The evaporator
    B. The condenser
    C. The dryer
    D. The compressor (B3.1)

53. When servicing the expansion valve, a technician can perform the following EXCEPT:
    A. adjusting the expansion valve using a torque wrench.
    B. adjusting the expansion valve using an Allen wrench.
    C. adjusting the expansion valve using a screwdriver.
    D. The expansion valve cannot be adjusted. (B3.6)

54. An ambient temperature sensor prevents the system from operating in low temperatures to prevent extremely low pressures in the suction side. Technician A says this is a zener diode that measures temperature change. Technician B says that in most A/C systems, a thermistor is supplied with V-Bat (battery voltage) and returns a portion of it as a signal. Who is correct?
    A. A only
    B. B only
    C. Both A and B
    D. Neither A nor B (D1.4)

55. To remove a mildew odor from the A/C output air:
    A. clean the evaporator with a disinfectant and water solution, and rinse it thoroughly.
    B. spray a disinfectant into the panel outlets.
    C. pour a small amount of alcohol into the air intake plenum.
    D. place an automotive deodorizer under the dash. (B3.8)

56. Technician A says that the refrigerant containers for R-22 and R-134a are color coded. Technician B says that the R-134a containers use 1/2-inch, 16 ACME threads, which cannot be mistakenly connected to an R-22 gauge set or recovery station. Who is correct?
    A. A only
    B. B only
    C. Both A and B
    D. Neither A nor B (B1.2)

57. Airflow through the A/C condenser will be significantly affected by all of the following EXCEPT:
    A. debris trapped in the fins.
    B. relative humidity of the outside air.
    C. bent cooling fins.
    D. vehicle speed. (B3.3)

58. To avoid mixing refrigerants, Technician A says that the R-134a service hose fittings are different from R-12 fittings. Technician B says that R-12 containers may be either white or gray with a yellow top, but R-134a containers are sky blue. Who is correct?
    A. A only
    B. B only
    C. Both A and B
    D. Neither A nor B (E2)

59. When evacuating an A/C system, which manifold gauge hose is connected to the vacuum pump?
    A. The high-pressure hose
    B. The low-pressure hose
    C. The center service hose
    D. Any hose (B1.6)

60. Manual temperature control systems may use a coolant control valve that is operated using any of the following EXCEPT:
    A. chassis air pressure.
    B. vacuum.
    C. a cable.
    D. a magneto. (D2.3)

61. During a performance test, the A/C compressor clutch is observed to be slipping. Technician A says that the air gap was probably set improperly. Technician B says that the pulley needs to be resurfaced. Who is correct?
    A. A only
    B. B only
    C. Both A and B
    D. Neither A nor B (B2.4)

62. Which of the following tools can be used to measure the freeze protection of engine coolant?
    A. Litmus strip
    B. Manometer
    C. Refractometer
    D. Opacity meter (C3)

63. There is a growling or rumbling noise at the A/C compressor with the compressor clutch engaged or disengaged. Technician A says the compressor bearing is defective. Technician B says the clutch bearing is defective. Who is correct?
    A. A only
    B. B only
    C. Both A and B
    D. Neither A nor B (A1)

64. Technician A says that the low-pressure cutout switch prevents compressor operation if the refrigerant charge has been lost or the ambient temperature is too cold. Technician B says that the high-pressure cutoff switch turns off the compressor if the system pressure is too high. Who is correct?
    A. A only
    B. B only
    C. Both A and B
    D. Neither A nor B (D1.3)

65. An electric condenser fan motor can be controlled by any of the following EXCEPT:
    A. an electronic relay.
    B. an independent electronic module.
    C. a multiplexed electronic module.
    D. an air solenoid controller. (D1.5)

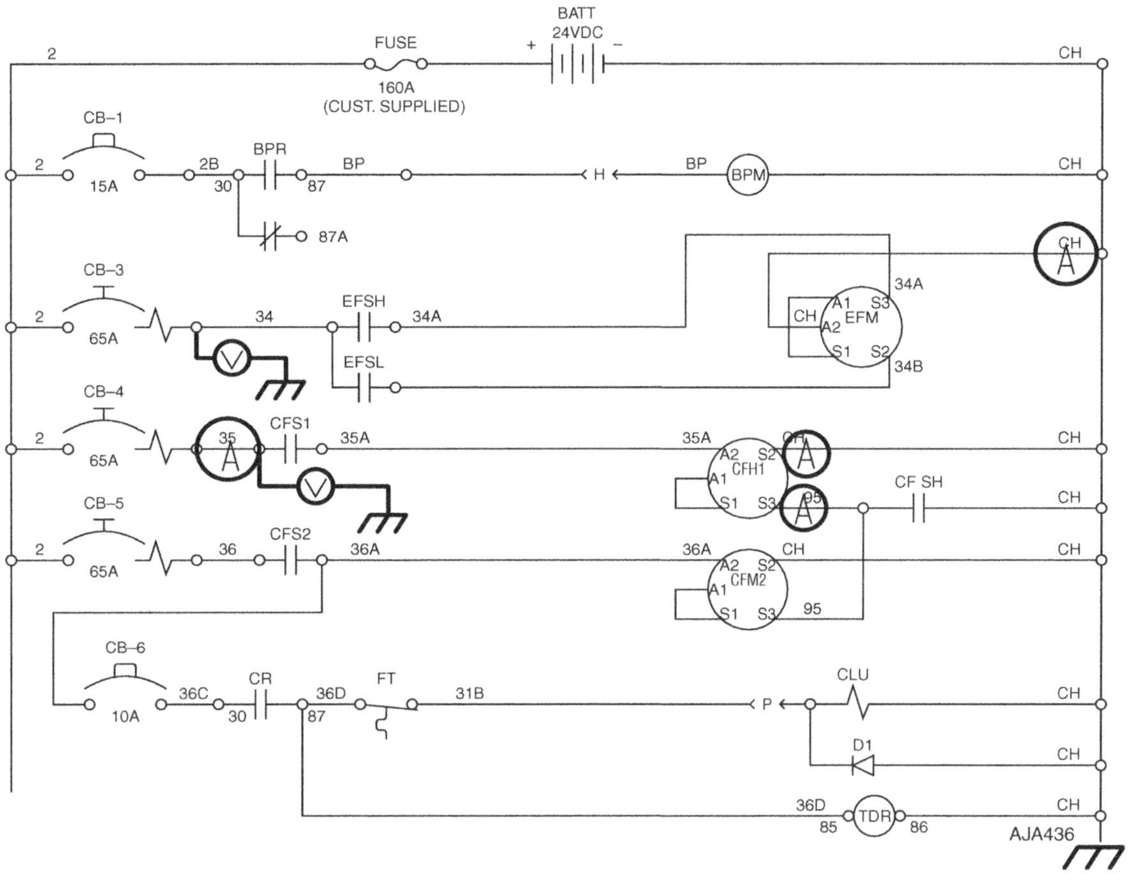

66. Refer to the figure. All of the following could prevent the A/C clutch from engaging EXCEPT:
    A. a faulty circuit breaker 1.
    B. ambient temperature below 40° F.
    C. the ambient temperature cutout switch stuck closed.
    D. open circuit in the 31B wire. (D1.1)

67. The auxiliary heater switch has a built-in LED to indicate that the system is on and to alert the operator or technician to any problems. The system is indicating a problem. The correct way to diagnose the system and repair it is to:
    A. turn the system off to eliminate the code.
    B. plug in a laptop computer to read the code.
    C. open the heater and perform routine maintenance.
    D. change the filter and nozzle. (C17)

6 Additional Test Questions for Practice

Additional Test Questions

Please note the letter and number in parentheses following each question. They match the task in Section 4 that discusses the relevant subject matter. You may want to refer to the overview using the cross-referencing key to help with questions posing problems for you.

1. Technician A says that the manifold gauge set is the only tool needed to diagnose HVAC systems. Technician B says a small thermometer is a valuable tool to evaluate the performance of an HVAC system. Who is correct?
   A. A only
   B. B only
   C. Both A and B
   D. Neither A nor B (A3)

2. An HVAC system controls all of the following EXCEPT:
   A. the engine radiator shutters.
   B. the blower motor speed.
   C. the fresh air damper actuator.
   D. the outside air door position. (D2.1)

3. What is the source of most non-condensable gases in refrigerant?
   A. Acid
   B. Air
   C. Moisture
   D. Oil (E5)

4. All of the following statements about nitrogen flushing the A/C system are true EXCEPT:
   A. the technician should install a pressure regulator on the supply tank.
   B. the technician should disconnect the A/C compressor.
   C. the technician should remove restrictive components (i.e., STV, TXV) from the system.
   D. nitrogen must not be allowed to escape into the atmosphere. (B1.7)

5. The water pump should be replaced any time:
   A. the fan clutch is replaced.
   B. the heater hoses are replaced.
   C. there is a leak from the weep hole.
   D. the thermostat sticks closed. (C6)

6. A bus HVAC is about to have the A/C system recharged. Which of these statements about the process is correct?
   A. The charging process is complete when the system reaches the correct evaporator temperature.
   B. When the low side no longer moves from a vacuum to a pressure, the process is complete.
   C. The bus engine must be running during recharging.
   D. Either a high-side or a low-side charging process can be used. (B1.8)

7. Which of the following is a characteristic of R-134a refrigerant?
   A. odorless
   B. a faint, ether-like odor
   C. a strong rotten egg odor
   D. a cabbage-like odor (B1.2)

8. All of these statements are true about a coolant control valve EXCEPT:
   A. it controls the flow of coolant through the heater core.
   B. it is part of the water pump.
   C. it may be cable operated.
   D. it may be electrically operated. (C10)

9. Which method of detecting refrigerant leaks is not recommended?
   A. Looking for bubbles after spraying with a soapy solution
   B. Using an electronic leak detector
   C. Using a flame-type leak detector
   D. Using a black light detector (B1.5)

10. Technician A says that you can test the A/C compressor clutch coil with an ohmmeter. Technician B says that connecting battery power to one terminal of the coil and grounding the other terminal can test the A/C compressor clutch coil. Who is correct?
    A. A only
    B. B only
    C. Both A and B
    D. Neither A nor B (B2.4)

11. Of the following, which is a normal range of low-side gauge operating pressures with R-134a?
    A. 5–10 psi
    B. 10–45 psi
    C. 60–80 psi
    D. 180–205 psi (B1.3)

12. Technician A says that you should replace the filter/dryer if the sight glass appears cloudy because it indicates a ruptured desiccant pack. Technician B says that you should only replace the filter/dryer if it has a leak. Who is correct?
    A. A only
    B. B only
    C. Both A and B
    D. Neither A nor B (B3.5)

13. Which of the following items is Most-Likely to cause elevated high-side pressure in an A/C system?
    A. Restricted airflow through condenser
    B. A thermostat stuck open
    C. A leaking thermal bulb
    D. Restriction in the filter dryer (B1.3)

14. When an expansion valve is properly installed, where should the thermal bulb and capillary tube be positioned?
    A. Fastened to the condenser fins using epoxy
    B. Located in the accumulator
    C. Held in contact with the evaporator outlet using insulating tape
    D. As an integral part of the orifice tube assembly (B3.6)

15. The test for non-condensable gases in recovered/recycled refrigerant involves:
    A. comparing the pressure of the recovered refrigerant in the container to the theoretical pressure of pure refrigerant at a given temperature.
    B. comparing the atmospheric pressure to the relative humidity.
    C. comparing the container pressure with the size of the container.
    D. testing the refrigerant with a halogen leak detector. (E5)

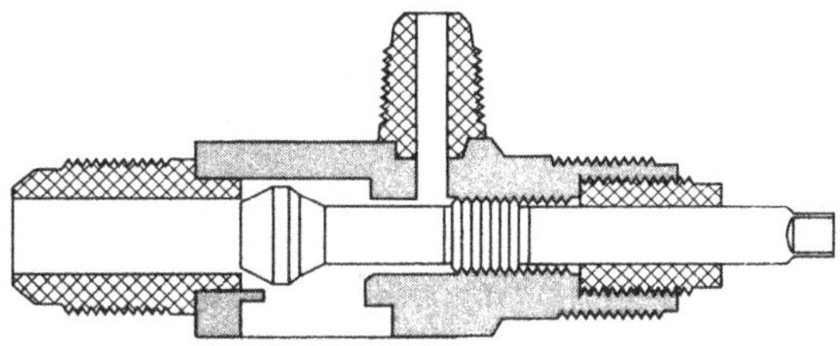

16. What is the position of the service valve shown in the figure?
    A. Front-seated position
    B. Back-seated position
    C. Mid-position
    D. Normal operating position (B3.9)

17. During an HVAC performance test, the technician notices that the A/C compressor outlet is nearly as hot as the upper radiator hose. Technician A says that this is a normal condition. Technician B says that the system is overcharged. Who is correct?
    A. A only
    B. B only
    C. Both A and B
    D. Neither A nor B (A2)

18. An A/C system with excessive high-side pressure could be the result of all of the following EXCEPT:
    A. an over-charge of refrigerant.
    B. a dirty condenser coil
    C. restricted airflow through the condenser.
    D. blockage of the expansion valve. (B1.3)

19. The A/C high-pressure switch is used to:
    A. boost the system high-side pressure.
    B. open the circuit to the A/C compressor clutch coil when the high-side pressure reaches its upper limit.
    C. ensure that system pressure remains at the upper limit.
    D. vent refrigerant from the compressor in the event of extremely high system pressure. (B2.2)

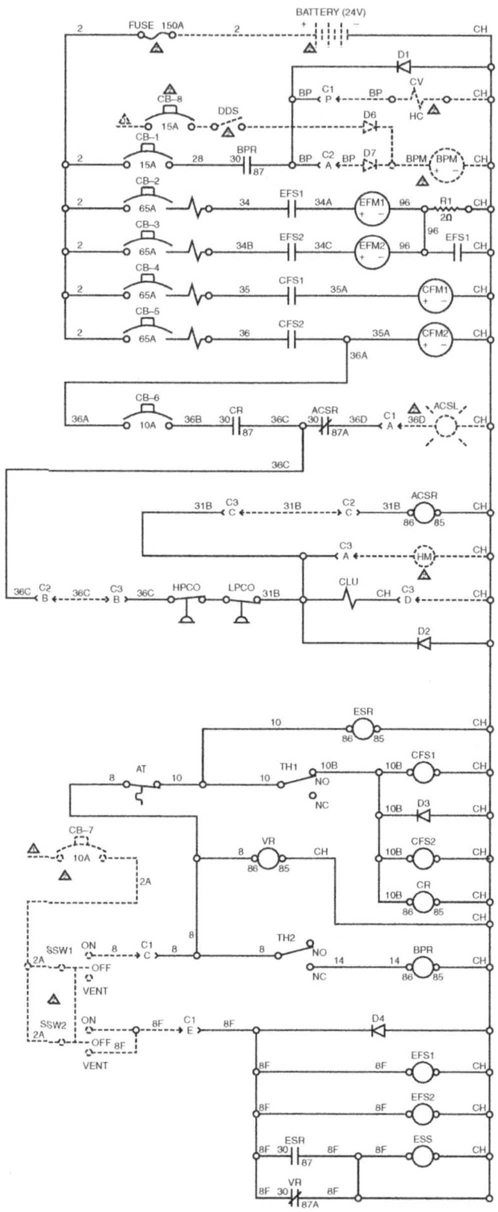

LEGEND

ACSR	AC STOP LIGHT RELAY
ACSL	AC STOP LIGHT
AT	AMBIENT THERMOSTAT SWITCH
BPM	BOOST PUMP RELAY
BPR	BOOST PUMP RELAY
C1	CONNECTOR 1
C2	CONNECTOR 2
C3	CONNECTOR 3
CB-1-8	CIRCUIT BREAKERS
CFM1	CONDENSER FAN MOTOR 1
CFM2	CONDENSER FAN MOTOR 2
CFS1	CONDENSER FAN SOLENOID 1
CFS2	CONDENSER FAN SOLENOID 2
CLU	COMPRESSOR CLUTCH
CR	CLUTCH RELAY
CV	ENGINE COOLANT VALVE (NC)
D1–D7	DIODES
DDSW	DRIVERS DEFROST SWITCH
EFM1,2	EVAPORATOR FAN MOTOR
EFS1	EVAPORATOR FAN SOLENOID 1
EFS2	EVAPORATOR FAN SOLENOID 2
ESR	EVAPORATOR SPEED RELAY
ESS	EVAPORATOR SPEED SOLENOID
HM	HOUR METER
HPCO	HIGH PRESSURE CUTOUT SWITCH
LPCO	LOW PRESSURE CUTOUT SWITCH
RJ	RESISTOR (.2 OHMS)
SSW1,2	SELECTOR SWITCH
TH1,2	THERMOSTAT SWITCH
VR	VENT RELAY

20. Refer to the figure. All of the following could prevent A/C compressor clutch engagement EXCEPT:
    A. a defective blower switch.
    B. an open low-pressure cutoff switch.
    C. a poor connection at the A/C thermostat.
    D. an open circuit breaker at CB 1. (B2.4)

21. Cooling and heating system hoses should be replaced for any of the following reasons EXCEPT:
    A. they leak at the hose clamps.
    B. they are cracked.
    C. they show signs of bulging.
    D. they feel spongy. (C4)

22. A faint hissing noise is heard from the area of the evaporator immediately after shutting down the engine with the compressor clutch engaged. Technician A says that the A/C system has a leak. Technician B says the thermal expansion valves are defective. Who is correct?
    A. A only
    B. B only
    C. Both A and B
    D. Neither A nor B (A1)

23. Technician A says that moisture in an air conditioning system can combine with the refrigerant to form harmful acids. Technician B says that moisture may be removed from an air conditioning system by evacuating. Who is correct?
    A. A only
    B. B only
    C. Both A and B
    D. Neither A nor B (B1.6)

24. Technician A says all A/C systems use a mineral-based internal lubricant. Technician B says POE oil is a nonsynthetic form of lubricant. Who is correct?
    A. A only
    B. B only
    C. Both A and B
    D. Neither A nor B (B1.9)

25. When replacing the heater core, foam tape is used:
    A. to insulate the heater core.
    B. to reduce noise from coolant surges.
    C. to protect and seal around the heater core.
    D. to seal the heater hose connections. (C11)

26. Which of these A/C gauge set readings would Most-Likely indicate that the A/C compressor reed valves are worn out?
    A. High-side pressure too low, and low-side pressure too high
    B. No pressure on either gauge
    C. Both gauges read pressure that is too low
    D. Both gauge readings are too high (B2.6)

27. The compressor discharge valve is designed to:
    A. open after vaporized refrigerant is compressed, allowing the refrigerant to move to the condenser.
    B. open before vaporized refrigerant is compressed, allowing the refrigerant to move to the evaporator.
    C. regulate system variable pressure.
    D. regulate A/C system temperature. (B2.6)

28. Technician A says that moisture can be removed from the A/C system after charging the system with new refrigerant. Technician B says moisture that enters the A/C circuit will be harmful to the system and cause poor performance. Who is correct?
    A. A only
    B. B only
    C. Both A and B
    D. Neither A nor B (B1.6)

29. Technician A says you can complete the high-side charging procedure with the engine running. Technician B says if liquid refrigerant enters the compressor, damage will result to the compressor. Who is correct?
A. A only
B. B only
C. Both A and B
D. Neither A nor B (B1.8)

30. Which of the following pressure combinations will the system show on a manifold gauge set if the expansion valve is not opening?
A. Low, low-side pressure and low, high-side pressure
B. Low, low-side pressure and high, high-side pressure
C. High, low-side pressure and high, high-side pressure
D. High, low-side pressure and low, high-side pressure (B1.3)

31. Technician A says a TXV not regulating properly could cause reduced airflow from the vent outlets. Technician B says a TXV not regulating properly could cause the evaporator to ice up. Who is correct?
A. A only
B. B only
C. Both A and B
D. Neither A nor B (B1.1)

32. Technician A says that the best way to repair wiring in the compressor clutch circuit is by soldering. Technician B says that twisting the wires together and securing them with electrical tape is sufficient. Who is correct?
A. A only
B. B only
C. Both A and B
D. Neither A nor B (D1.3)

33. The A/C compressor clutch will not engage in any mode. The clutch engages when a technician installs a jumper wire across the terminals of the low-pressure cutout switch connector. Technician A says that the low-pressure cutout switch must be defective. Technician B says that the refrigerant charge might be low. Who is correct?
A. A only
B. B only
C. Both A and B
D. Neither A nor B (B2.1)

34. Filters installed in the HVAC air delivery system:
A. are always made of fiberglass mesh to resist corrosion.
B. are designed to remove moisture from bus air.
C. are not individually replaceable.
D. are designed to remove dust and dirt from bus air. (B3.8)

35. With the selector lever in the High Cool position, an HVAC system outputs cold air for about 15 minutes, after which the output air becomes warm. The Most-Likely cause of this problem is:
A. a defective thermal expansion valve.
B. a defective coolant control valve.
C. a defective coolant pump.
D. a defective fresh air door. (A3)

36. A bus technician discovers a heater core leak in the driver heater that requires replacing the heater. Technician A says that it is important to add the proper amount of refrigeration oil before installation. Technician B says that a PAG-based lubricant is used in modern heater cores. Who is correct?
    A. A only
    B. B only
    C. Both A and B
    D. Neither A nor B (C11)

37. Which of the following statements about coolant control valve replacement in an ATC system is true?
    A. Refrigerant must be recovered before the coolant control valve is replaced.
    B. Heater hoses connected to the coolant control valve may be clamped during the replacement procedure to maintain coolant in the system.
    C. Access to the coolant control valve is gained through the fresh air door.
    D. The coolant control valve is an integral part of the heater hose. (D2.3)

38. Which sensor signal to the HVAC control module (ECM) would cause the A/C compressor **not** to disengage?
    A. Compressor discharge temperature (CTS) sensor
    B. Ambient air temperature (ATS) sensor.
    C. Water temperature sensor (WTS)
    D. Compressor discharge sensor (CDS) (D1.4)

39. When refilling a bus cooling system with premixed ELC (extended life coolant), which of the following is correct practice?
    A. Add coolant conditioner directly to the premix.
    B. Only add distilled water to the premix.
    C. Top up only with propylene glycol (PG) concentrate.
    D. Never add anything to the ELC premix. (C8)

40. Technician A says that according to new environmental laws, shut-off valves must be placed in the closed position every time the A/C system is turned off. Technician B says that according to new environmental laws, shut-off valves must be located no more than 12 inches from test hose service ends. Who is correct?
    A. A only
    B. B only
    C. Both A and B
    D. Neither A nor B (E1)

41. Which of the following modes will provide the **least** effective windshield defrosting in the defrost mode?
    A. 50 percent of the output air is directed to the floor and 50 percent to the windshield.
    B. 15 percent of the output air is directed to the floor and 85 percent to the windshield.
    C. 25 percent of the output air is directed to the floor and 75 percent to the windshield.
    D. 30 percent of the output air is directed to the floor and 70 percent to the windshield. (A3)

42. When checking the coolant condition with an SCA (supplemental coolant additive) test strip, the technician finds that the coolant is severely over-conditioned. What should the technician do?
    A. Add more antifreeze to increase the SCA.
    B. Continue to run the truck until the next PMI.
    C. Drain the entire coolant system and add the proper SCA mixture.
    D. Run the truck with no SCA additives until the next PMI. (C3)

43. A driver report indicates that the driver-side defroster operates only in the high-speed mode. The first step to determine the possible reason for this is to:
    A. change the switch.
    B. change the motor.
    C. replace the resistor for low speed.
    D. check the schematic to determine what controls the motor speed. (C13)

44. A screen is located in the expansion valve of an A/C system. Technician A says that the screen is a filter used to prevent particulate from circulating through the system. Technician B says that the screen is used to improve atomization of the refrigerant. Who is correct?
    A. A only
    B. B only
    C. Both A and B
    D. Neither A nor B (B3.7)

45. To detect a refrigerant leak, hold the leak detector sensor:
    A. within three inches of the fitting.
    B. just below the fitting.
    C. right next to the fitting.
    D. just above the fitting. (B1.5)

46. Technician A says that the HVAC systems on most buses with diesel engines operate totally independently from the engine control system. Technician B says that most of the HVAC systems on these vehicles provide inputs and receive outputs from the engine control unit. Who is correct?
    A. A only
    B. B only
    C. Both A and B
    D. Neither A nor B (D1.4)

47. All of these statements about cooling system service are true EXCEPT:
    A. when the cooling system pressure is increased, the boiling point increases.
    B. the boiling point decreases when more antifreeze is added to the coolant.
    C. good quality ethylene glycol antifreeze contains antirust inhibitors.
    D. coolant solutions should be recovered, recycled, and handled as hazardous waste. (C8)

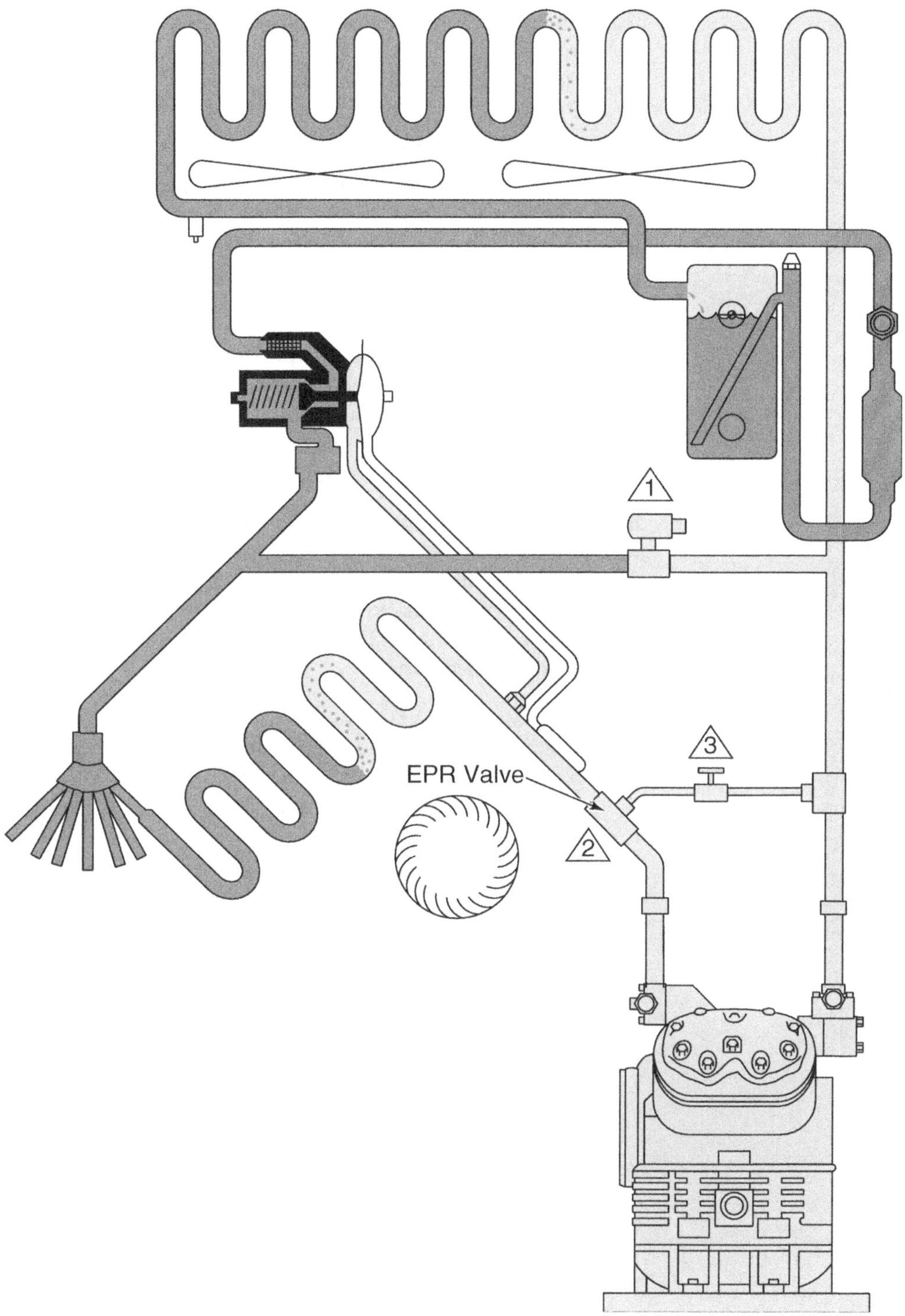

48. If the component indicated by an arrow in the figure was not used, what would be the A/C performance result?
    A. Higher A/C high-side pressure
    B. Improperly filtered refrigerant
    C. Evaporator icing
    D. Noisier A/C operation (A3)

49. A cooling system pressure tester can be used to test:
    A. thermostats.
    B. radiators, pressure caps, and hoses.
    C. A/C leaks.
    D. the fresh air damper door actuator. (C5)

50. The A/C compressor drive belt should be adjusted:
    A. using a belt tension gauge.
    B. so there is no deflection at maximum engine speed.
    C. so there is no static deflection.
    D. as tightly as possible. (B2.3)

51. Which of the following is **not** a type of bus HVAC system?
    A. Mechanical-thermostat controlled
    B. Microprocessor temperature control
    C. Recirculating chilled and heated water
    D. Electronic temperature control (ETC) (A3)

52. Bus A/C systems using R-22 have flared and threaded service ports. Service fittings on systems with R-134a use:
    A. compression fittings.
    B. SAE-approved quick-connect couplings.
    C. the same fittings as R-22 systems.
    D. 10-mm threaded ports. (A3)

53. While checking an ATC system with a laptop computer and specific program for the equipment, the technician sees the code for return air sensor failure. The technician should:
    A. look for an open wire.
    B. look for a short circuit to ground.
    C. test the control module.
    D. look for a bad mode selector. (D2.6)

54. Technician A says that as specified by SAE J1991, recycled refrigerant cannot contain more than 15 ppm moisture contaminants by weight. Technician B says that as specified by SAE J1991, recycled refrigerant cannot contain more than 330 ppm non-condensable gases (air) by weight. Who is correct?
    A. A only
    B. B only
    C. Both A and B
    D. Neither A nor B (E3)

55. A routine A/C maintenance service should include all of the following EXCEPT:
    A. tightening the condenser lines.
    B. removing debris from the condenser fins.
    C. straightening the condenser fins.
    D. checking the condenser mounts. (B3.4)

56. Which of the following statements is true about checking the A/C compressor lubricant level?
    A. The A/C system must first be evacuated for at least 30 minutes.
    B. The system must be operated for at least 15 to 20 minutes and stabilized before checking the compressor lubricant level.
    C. The oil level should be up to the full mark on the dipstick.
    D. The old lubricant must be measured before the new lubricant is added. (B2.5)

57. Technician A says that you should test the coolant condition before replacing the coolant conditioner cartridge. Technician B says that the coolant conditioner cartridge should be replaced every time the engine lube and fuel filters are changed. Who is correct?
    A. A only
    B. B only
    C. Both A and B
    D. Neither A nor B (C3)

58. Technician A says that the compressor oil needs to be checked when there is evidence of loss of system oil. Technician B says that when replacing refrigerant oil, it is important to use the specific type and quantity of oil recommended by the compressor manufacturer. Who is correct?
    A. A only
    B. B only
    C. Both A and B
    D. Neither A nor B (B1.9)

59. The thermal bulb on an expansion valve must be installed in contact with:
    A. the evaporator outlet tube.
    B. the condenser fins.
    C. the suction hose.
    D. the refrigerant. (B3.6)

60. Poor cooling from the A/C system that uses a TXV can be caused by all of the following EXCEPT:
    A. the fan clutch always being engaged.
    B. a coolant valve that is stuck open.
    C. a low refrigerant charge.
    D. a refrigerant overcharge. (B1.1)

61. On a cycling clutch A/C system, the low-side reading is too high and the high-side reading is too low. Technician A says that the compressor may have a faulty reed valve. Technician B says that an overcharge of refrigerant oil is a possible cause. Who is correct?
    A. A only
    B. B only
    C. Both A and B
    D. Neither A nor B (B1.3)

62. A/C system pressures vary with all of the following EXCEPT:
    A. altitude.
    B. ambient temperature.
    C. ambient air humidity
    D. cab humidity. (B1.3)

63. Technician A says that a clogged heater core could cause insufficient heater output. Technician B says that an improperly adjusted coolant control valve could cause insufficient heater output. Who is correct?
    A. A only
    B. B only
    C. Both A and B
    D. Neither A nor B (C1)

64. The A/C condenser has several bent fins and a moderate accumulation of dead insects. Technician A says that the condenser should be replaced or it will cause the high-pressure relief valve to trip. Technician B says that the condenser fins should be straightened and the dead insects removed to optimize A/C system performance. Who is correct?
A. A only
B. B only
C. Both A and B
D. Neither A nor B (B3.3)

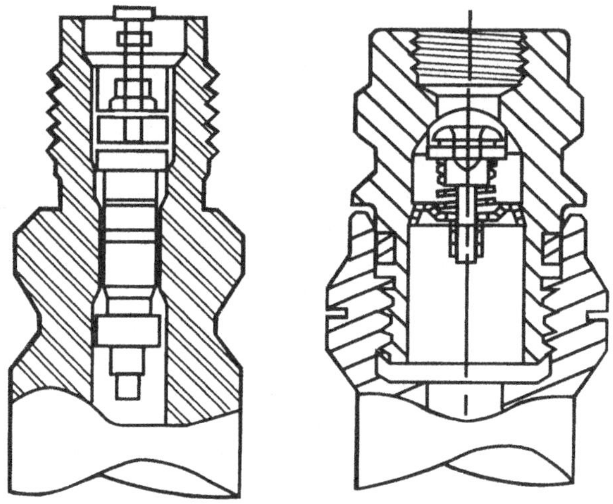

A B

65. Technician A says that a system with a service port fitting shown in figure A is designed to use R-134a as a refrigerant. Technician B says that the service port fitting shown in figure B is for use with R-12. Who is correct?
A. A only
B. B only
C. Both A and B
D. Neither A nor B (A3)

66. Technician A says that systems use an accumulator to evaporate any liquid refrigerant before it enters the compressor. Technician B says that systems that use a TXV have a receiver tank. Who is correct?
A. A only
B. B only
C. Both A and B
D. Neither A nor B (A3)

67. When evacuating an A/C system, the vacuum pump should be operated to achieve a minimum of:
A. 2500 microns.
B. 1200 microns.
C. 500 microns.
D. 5000 microns. (B1.6)

68. Technician A says that the A/C compressor high-pressure relief valve must be replaced if it operates due to extreme pressure in the A/C system. Technician B says that the compressor must be replaced if the high-pressure relief valve operates. Who is correct?
    A. A only
    B. B only
    C. Both A and B
    D. Neither A nor B (B3.10)

69. Technician A says that in a typical bus A/C system, refrigerant enters the compressor as a low-pressure gas. Technician B says that in an A/C circuit, refrigerant is designed to boil in the condenser heat exchanger. Who is correct?
    A. A only
    B. B only
    C. Both A and B
    D. Neither A nor B (A3)

70. The inside of the windshield has a sticky film. Technician A says to check the engine coolant level. Technician B says the heater core may be leaking. Who is correct?
    A. A only
    B. B only
    C. Both A and B
    D. Neither A nor B (C2)

71. Technician A says that when evacuating a system, the vacuum pump must run until 1200 microns of vacuum to remove moisture from the system. Technician B says that observing the action of the gauges while the system is being evacuated may help indicate the presence of either a blockage or a leak. Who is correct?
    A. A only
    B. B only
    C. Both A and B
    D. Neither A nor B (B1.6)

72. In a microprocessor-controlled A/C system, the temperature control is set at 70° F (21° C), and the bus temperature is 80° F (27° C) after driving one hour. All refrigerant pressures are normal. Technician A says the bus sensor may be defective. Technician B says the fresh air damper may be sticking. Who is correct?
    A. A only
    B. B only
    C. Both A and B
    D. Neither A nor B (D2.1)

73. Technician A says some manufacturers recommend the installation of an inline filter between the evaporator and the compressor as a method for cleanup. Technician B says a bypass oil filter on the compressor oil supply is adequate for cleanup. Who is correct?
    A. A only
    B. B only
    C. Both A and B
    D. Neither A nor B (B1.7)

74. If R-22 comes into contact with a flame:
    A. it will explode.
    B. it will form a nontoxic gas.
    C. it will form chlorine crystals.
    D. it will form phosgene gas. (B1.5)

75. While conducting a performance test on a driver heater, a technician finds that only a small amount of air is directed to the windshield in defrost mode. The Most-Likely cause of this problem is:
    A. a defective microprocessor.
    B. a defective blend door actuator.
    C. an open blower motor resistor.
    D. an improperly adjusted mode door cable (A3)

76. Which of the following is a typical bus cooling system operating pressure?
    A. 1 psi
    B. 4 psi
    C. 20 psi
    D. 30 psi (C5)

77. Which of the following tests or instruments would provide the most accurate measurement of coolant antifreeze protection?
    A. Hydrometer
    B. SCA test strip
    C. Litmus test
    D. Refractometer (C3)

78. Technician A says that some systems have an inline filter. Technician B says that petroleum jelly should be used to lubricate new A/C O-rings. Who is correct?
    A. A only
    B. B only
    C. Both A and B
    D. Neither A nor B (B3.2)

79. The A/C compressor high-pressure relief valve:
    A. is calibrated by shimming it to the proper depth.
    B. must be replaced if it ever vents refrigerant from the system.
    C. will reset itself when A/C system pressure returns to a safe level.
    D. is not used in R-134a systems. (B3.10)

80. In an ATC system, a banging noise comes from the roof as the temperature set point is reached. The Most-Likely cause of this problem is:
    A. loose evaporator motor.
    B. an open access panel.
    C. a coolant valve shutting and a water hammer occurring
    D. arcing in the evaporator motor. (D2.4)

81. What is the function of the expansion valve?
    A. It allows water to drain from the evaporator case.
    B. It allows refrigerant to be metered into the evaporator.
    C. It regulates refrigerant flow through the condenser.
    D. It regulates airflow through the evaporator. (B3.7)

82. An electric condenser fan may be controlled by any of the following EXCEPT:
    A. the motor control unit.
    B. the HVAC ECM.
    C. a manual switch.
    D. an electronic relay. (D1.5)

83. Which of the following would be the most common maintenance problem in bus heating systems?
    A. Clogged heater cores
    B. Defective control units
    C. Faulty control valves
    D. Coolant leaks (C3)

84. To check for cooling system external leaks, pressurize the system and:
    A. use a black light to find leaks.
    B. use a leak detector to find leaks.
    C. perform a visual inspection.
    D. watch for white smoke. (C3)

85. Flushing the cooling system does not:
    A. remove rust from the system.
    B. remove contaminants from the system.
    C. increase the life of the cooling system.
    D. remove acids of combustion from the cooling system. (C8)

86. Technician A says that the best way to ensure that an A/C compressor has the proper amount of lubricant is to drain the compressor and add lubricant to the manufacturer's specifications. Technician B says that if you have a doubt about the lubricant level, you should add an oil charge to the system. Who is correct?
    A. A only
    B. B only
    C. Both A and B
    D. Neither A nor B (B2.5)

87. During an HVAC performance test, the technician hears the A/C compressor clutch slip briefly upon engagement. The Most-Likely cause is:
    A. a worn out compressor clutch coil.
    B. a defective A/C compressor clutch relay.
    C. the compressor clutch air gap is too large.
    D. the compressor clutch bearing is worn. (A1)

88. Which of the following is LEAST-Likely to cause windshield fogging in the DEFROST mode?
    A. A leaking heater core
    B. A clogged evaporator drain
    C. A water leak into the plenum chamber
    D. Moisture in the refrigerant (C2)

89. Before replacing an HVAC electronic control panel, the technician should:
    A. remove the control cables from the vehicle.
    B. disconnect the batteries.
    C. disassemble the dash panel.
    D. apply dielectric grease to the switch contacts. (D1.6)

90. The expansion valve is located at:
    A. the inlet line of the evaporator.
    B. the outlet line of the evaporator.
    C. the inlet line of the compressor.
    D. the outlet line of the condenser. (B3.6)

91. Which of these terms correctly describes refrigerant that has been removed from a system and stored in an external container?
    A. Recycled
    B. Recovered
    C. Reclaimed
    D. Refined (E3)

92. Technician A says that anyone who purchases R-134a must maintain records for three years indicating the name and address of the supplier. Technician B says the supplier must maintain sales records of refrigerant purchases. Who is correct?
    A. A only
    B. B only
    C. Both A and B
    D. Neither A nor B (E1)

93. When diagnosing a computer-controlled ATC A/C system, a diagnostic trouble code is displayed indicating a fault in the fresh air damper feedback. Technician A says the first step in the repair process is to replace the fresh air damper actuator. Technician B says you need to check the harness and connectors. Who is correct?
    A. A only
    B. B only
    C. Both A and B
    D. Neither A nor B (D2.4)

94. In high cool A/C mode:
    A. the blowers are operating at high speed.
    B. the blowers operate at low speed.
    C. the compressor clutch cannot disengage.
    D. the blower is disabled. (B1.1)

95. Technician A says you check the calibration accuracy of an HVAC system using A/C pressure gauges only. Technician B says you can recalibrate all microprocessor systems in the field. Who is correct?
    A. A only
    B. B only
    C. Both A and B
    D. Neither A nor B (D2.1)

96. Any of the following can be used to clean road debris from the condenser fins EXCEPT:
    A. a mild saline solution.
    B. a soft whisk broom.
    C. compressed air.
    D. a mild soap and water solution. (B3.3)

97. Before disposing of an empty or near empty original container that was used to ship refrigerant from the factory, you should perform which of the following?
    A. Clean it and keep it for storage of recycled refrigerant.
    B. Open the valve completely and paint an X on the cylinder.
    C. Flush it with oil and nitrogen to keep it from rusting.
    D. Recover remaining refrigerant, evacuate the cylinder, and mark it empty. (E4)

98. Technician A says that the best way to test an electrically operated coolant control valve is to disconnect the power and observe whether coolant flows through it. Technician B says an electrically operated coolant control valve can be tested by applying current and checking if the valve operates. Who is correct?
    A. A only
    B. B only
    C. Both A and B
    D. Neither A nor B (C10)

99. Technician A says most fully automatic temperature control systems have some form of self-diagnostic program that will display trouble codes. Technician B says that, depending on the system design, trouble codes may be displayed digitally on the control assembly or on a hand-held scan tool. Who is correct?
    A. A only
    B. B only
    C. Both A and B
    D. Neither A nor B (D2.1)

100. The LEAST-Likely cause of the high-pressure relief valve tripping open is:
    A. improper radiator fan operation.
    B. a clogged condenser.
    C. inoperative high-pressure cutout.
    D. a defective A/C compressor. (B2.1)

101. Which type of heating system is the most common type used in buses?
    A. Forced air convection heater
    B. Immersion heater
    C. Fuel-fired heater
    D. Electric heater (A3)

102. Each of these is an indication that the thermostat has opened EXCEPT:
    A. visible coolant flow in the surge tank.
    B. the upper radiator hose is hot.
    C. the lower radiator hose is hot.
    D. the temperature gauge has stabilized in the normal range. (C7)

103. The refrigerant line leading from the evaporator to the compressor contains refrigerant as a:
    A. low-pressure gas.
    B. low-pressure liquid.
    C. high-pressure gas.
    D. high-pressure liquid. (A3)

104. Technician A says that some ATC systems use a microprocessor built into the bus PLC. Technician B says that some ATC systems use a microprocessor that can be replaced independently from the PLC. Who is correct?
    A. A only
    B. B only
    C. Both A and B
    D. Neither A nor B (D2.6)

105. The easiest way to check the calibration of an ATC system is to:
    A. use an ohmmeter to measure the resistance of the ATC sensor.
    B. measure the voltage drop across the ATC sensor.
    C. turn the calibration screw on the ATC control unit.
    D. use a thermometer to measure the cab temperature and compare the actual temperature to the return air sensor temperature reading. (D2.5)

106. Technician A says the easiest way to identify the type of refrigerant for a system is to check the unit specifications label. Technician B says that R-134a is the only refrigerant that may be vented to the atmosphere. Who is correct?
    A. A only
    B. B only
    C. Both A and B
    D. Neither A nor B (B1.2)

107. Coolant conditioner performs all of the following tasks EXCEPT:
    A. raising the boiling point of the coolant.
    B. inhibiting rust and debris from the coolant.
    C. lubricating the cooling system internally.
    D. preventing cavitation corrosion of wet cylinder liners. (C3)

108. Oil and dirt accumulation on an A/C hose connection may indicate:
    A. excessive pressure in the system.
    B. a refrigerant leak.
    C. a defective compressor shaft seal.
    D. that there is too much oil in the system. (B1.4)

109. When attempting to verify a leaking evaporator core, the technician is LEAST-Likely to sense refrigerant with the detector probe:
    A. at the evaporator housing drain.
    B. over the expansion valve.
    C. at the overhead vents.
    D. at the defroster vents. (B3.7)

110. An ATC control panel is suspected of causing an A/C performance problem in a bus. Technician A says to replace the ATC control computer. Technician B says to perform the sequential diagnostic steps in the HVAC manufacturer's service manual before replacing components. Who is correct?
    A. A only
    B. B only
    C. Both A and B
    D. Neither A nor B (D2.5)

111. A compressor cycling on and off too fast is most commonly caused by:
    A. a defective compressor clutch.
    B. a defective control switch.
    C. an overcharged system.
    D. a low refrigerant charge. (B1.4)

112. Technician A says a system having high, low-side pressure accompanied by a continuously running compressor indicates that the expansion valve is stuck open. Technician B says a system having a high, low-side reading accompanied by a continuously running compressor indicates a problem with the compressor clutch. Who is correct?
    A. A only
    B. B only
    C. Both A and B
    D. Neither A nor B (B1.3)

113. To prevent overfilling of recovery cylinders, the service technician must:
    A. monitor cylinder pressure as the cylinder is being filled.
    B. monitor cylinder weight as the cylinder is being filled.
    C. make sure cylinder safety relief valves are in place and operational.
    D. occasionally shake the cylinder and observe any change of pressure while filling. (E2)

114. All of the statements about replacing an HVAC control panel are true EXCEPT:
    A. you remove the negative battery cable before control panel service.
    B. you must recover the refrigerant before removing the control panel.
    C. if the bus contains a PLC system, wait the specified period after you remove the negative battery cable.
    D. self-diagnostic tests may indicate a defective control panel. (D1.6)

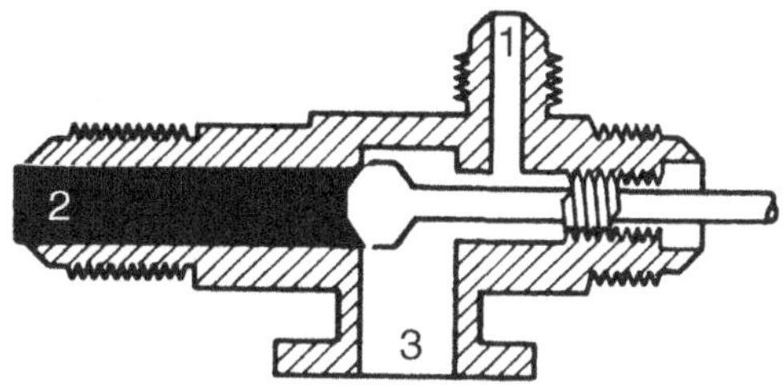

Front-Seated Position

115. With a stem-type service valve in the position shown in the figure:
    A. you can diagnose the refrigerant system with a manifold gauge set.
    B. the refrigerant system operates normally with no pressure at the gauge ports.
    C. the refrigerant system is isolated from the compressor for compressor removal.
    D. the refrigerant system may be discharged, evacuated, and recharged. (B3.9)

116. When pressure-testing a cooling system, there are no obvious external leaks but the system cannot maintain pressure. The Most-Likely cause of this problem is:
    A. a leaking evaporator.
    B. a defective heater valve.
    C. a stuck-open thermostat.
    D. a blown head gasket in the engine. (C3)

117. Never operate the compressor with the high-side service valve:
    A. back-seated.
    B. front-seated.
    C. in mid-position.
    D. hot. (B3.9)

7 Appendices

Answers to the Test Questions for the Sample Test Section 5

1. C	18. A	35. C	52. D
2. A	19. C	36. A	53. D
3. C	20. B	37. D	54. A
4. C	21. B	38. D	55. A
5. C	22. C	39. D	56. C
6. C	23. A	40. D	57. B
7. B	24. C	41. A	58. C
8. B	25. A	42. A	59. C
9. C	26. B	43. A	60. D
10. C	27. A	44. B	61. A
11. C	28. A	45. A	62. C
12. A	29. A	46. C	63. D
13. B	30. A	47. C	64. C
14. C	31. A	48. D	65. D
15. C	32. B	49. B	66. A
16. D	33. A	50. C	67. B
17. B	34. C	51. A	

Explanations to the Answers for the Sample Test Section 5

Question #1

Answer C is correct. A technician should never attempt to charge an A/C system through the high-side service valve when the engine is running; this high-pressure will likely exit the vehicle and enter your charging tank and could explode.

Answer A is incorrect. A technician can install refrigerant through both service valves when the engine is not running.

Answer B is incorrect. A technician can install refrigerant through the low-side service valve when the engine is running.

Answer D is incorrect. A technician can charge an A/C system directly from an approved charging station.

Question #2

Answer A is correct. Only Technician A is correct. A growling noise from the water pump indicates a worn bearing.

Answer B is incorrect. Cavitation damage will not cause this symptom. Poor circulation due to the lack of water circulation will result from a damaged water pump impeller.

Answer C is incorrect. Only Technician A is correct.

Answer D is incorrect. Only Technician A is correct.

Question #3

Answer C is correct. Both Technicians A and B are correct. Excessive high-side pressure that caused the pressure relief valve to operate might have been the result of over pressure. A faulty high-pressure relief valve could also cause the discharge.

Answer A is incorrect. Technician B is also correct.

Answer B is incorrect. Technician A is also correct.

Answer D is incorrect. Both technicians are correct.

Question #4

Answer C is correct. Both Technicians A and B are correct. A low heat problem could be caused by a misadjusted temperature control or by a clogged heater core.

Answer A is incorrect. Technician B is also correct.

Answer B is incorrect. Technician A is also correct.

Answer D is incorrect. Both technicians are correct.

Question #5

Answer C is correct. A 27 in. Hg of vacuum is required to remove all moisture from the container.

Answer A is incorrect. 20 in. Hg is not sufficient to ensure that all moisture has been removed from the tank.

Answer B is incorrect. 22 in. Hg is not sufficient to ensure that all moisture has been removed from the tank.

Answer D is incorrect. 12 in. Hg is not sufficient to ensure that all moisture has been removed from the tank.

Question #6
Answer C is correct. A suction filter is the best method to remove debris and acid from a system.

Answer A is incorrect. AC flush solvent is not recommended due to the size of the unit and the many tubes in the coils where the flushing agent could become trapped. Condensers that are used on R-134a systems have very small passages. In the event of a mechanical failure, if the passages of the condenser get clogged, the condenser needs to be replaced.

Answer B is incorrect. CFC refrigerants must never be used to flush the A/C system.

Answer D is incorrect. Nitrogen flushing will not clean residue from the lines and hoses.

Question #7
Answer B is correct. Only Technician B is correct. Deformed or improperly aligned condenser mounting insulators could damage the condenser and refrigerant lines. The insulators are made from rubber and act to cushion the vibration of the bus from the condenser and lines. If the insulators are deformed or misaligned, leaks could develop in these areas.

Answer A is incorrect. Damaged or misaligned condenser mounts should not affect compressor operation.

Answer C is incorrect. Only Technician B is correct.

Answer D is incorrect. Only Technician B is correct.

Question #8
Answer B is correct. Only Technician B is correct. The odor of leaking antifreeze can be drawn from the heater core even when operating in the cool mode. If the heater core was causing the odor, the technician would smell a sweet aroma with the HVAC controls set in the cool mode. Other signs of a leaking heater core might be coolant in the evaporator drains, low coolant in the surge tank, or a low coolant light on the dash.

Answer A is incorrect. R-134a is an odorless gas and would not cause a sweet odor.

Answer C is incorrect. Only Technician B is correct.

Answer D is incorrect. Only Technician B is correct.

Question #9
Answer C is correct. Both Technicians A and B are correct. When a compressor is replaced, the dryer should be replaced. Also, whenever the A/C system has been open to the atmosphere, the dryer should be replaced. When replacing the dryer, it is important to mount it last to prevent it from being exposed to the atmosphere for an extended period of time.

Answer A is incorrect. Technician B is also correct.

Answer B is incorrect. Technician A is also correct.

Answer D is incorrect. Both technicians are correct.

Question #10
Answer C is correct. Both Technicians A and B are correct. When installing a new or rebuilt compressor, you should first turn it over by hand and drain any oil shipped with the compressor, then add the correct amount of the right oil before installing it. Some compressors are shipped without oil. Most manufacturers recommend draining the old compressor into a measuring cup. If less than 2 ounces is drained from the old compressor, then 2 ounces of new oil should be added to the new compressor. If 2 ounces of oil or more is drained from the old compressor, then the same amount of new oil should be added to the new compressor. After adding the oil, the compressor should be turned by hand to help displace the oil to prevent damage when the compressor is first turned on.

Answer A is incorrect. Technician B is also correct.

Answer B is incorrect. Technician A is also correct.

Answer D is incorrect. Both technicians are correct.

Question #11
Answer C is correct. The recycling process reduces contaminants used in refrigerant by using oil separation and filter core dryers.

Answer A is incorrect. Restoration is a term generally applied to old buildings.

Answer B is incorrect. Recovery is the process by which refrigerant is removed from a system.

Answer D is incorrect. Reclamation is an industrial process by which refrigerant is restored to its original condition by an outside source.

Question #12
Answer A is correct. Only Technician A is correct. When the compressor clutch is not engaged, the pressure in the A/C system equalizes. Another possible cause for the pressures to be the same is if the manifold valves are in the open position. When checking pressures, it is important to have the manifold valves closed.

Answer B is incorrect. A restriction in the expansion valve will cause low, but not identical, system pressures.

Answer C is incorrect. Only Technician A is correct.

Answer D is incorrect. Only Technician A is correct.

Question #13
Answer B is correct. Only Technician B is correct. The heater control valves may be cable operated. Other methods to operate heater control valves are electrical solenoids.

Answer A is incorrect. The heater control valves are not operated by engine speed.

Answer C is incorrect. Only Technician B is correct.

Answer D is incorrect. Only Technician B is correct.

Question #14
Answer C is correct. The relationship of pressure to temperature at a constant volume is direct, so that at 70° F the pressure is about 70 psi. This is known as Charles' Gas Law.

Answer A is incorrect. If a cylinder had 220 psi of pressure at 70° F, then it would definitely have some other chemical mixed with it. A technician would need to use an identifier to find out what else is mixed with the refrigerant.

Answer B is incorrect. 125 psi of pressure in a cylinder of refrigerant at 70° F would indicate a high percentage of air. The refrigerant would need to be recovered and recycled to remove the air.

Answer D is incorrect. The relationship of pressure to temperature at a constant volume is direct, so that at 70° F the pressure is about 70 psi, not 30 psi.

Question #15
Answer C is correct. An improperly set compressor clutch air gap will not affect belt alignment.

Answer A is incorrect. A bent mounting bracket can cause belt misalignment.

Answer B is incorrect. A faulty clutch bearing can cause belt misalignment.

Answer D is incorrect. A faulty idler pulley will cause belt misalignment.

Question #16
Answer D is correct. The only accurate way to measure the amount of oil in a compressor is to observe the sight glass while the system is operating and stabilized.

Answer A is incorrect. This is not an acceptable method of adjusting the lubricant level in the system.

Answer B is incorrect. This is not an acceptable method because the A/C system must first be recovered and the compressor removed from the vehicle.

Answer C is incorrect. This method is not correct.

Question #17
Answer B is correct. A cracked mounting plate will not cause internal compressor damage.

Answer A is incorrect. A cracked mounting plate could cause drive belt wear by not holding the compressor in place. This would misalign the compressor and cause the belt to wear.

Answer C is incorrect. A cracked mounting plate can cause a vibration with the compressor clutch engaged. This vibration can appear to be caused by a faulty compressor, but the cause is just a cracked mounting plate.

Answer D is incorrect. A cracked mounting plate can cause belt squeal or chatter because of misaligned pulleys.

Question #18
Answer A is correct. Only Technician A is correct. R-22 A/C service valves have a replaceable Schrader valve.

Answer B is incorrect. R-134a service valves cannot be rear-seated.

Answer C is incorrect. Only Technician A is correct.

Answer D is incorrect. Only Technician A is correct.

Question #19
Answer C is correct. The expansion separates the high and low side of the refrigeration system. Refrigerant enters the expansion valve as a high-pressure liquid. After passing through the needle and seat, the refrigerant is at a much lower pressure, so it starts to boil. This boiling allows the refrigerant to take on latent heat from the surface around it in the evaporator. The refrigerant continues to boil until it completely turns into a vapor and is routed to the compressor inlet.

Answer A is incorrect. The vapor line (also called the suction line) carries low-pressure vapor to the compressor inlet connection.

Answer B is incorrect. The condenser removes heat from the high-pressure gas, which allows it to condense back into a liquid.

Answer D is incorrect. The capillary tube is a temperature-sensing device on a TXV.

Question #20
Answer B is correct. A restriction in the high-pressure hose will result in a band of frost on the high-pressure line. All of the components in the high side of the system should be hot while the system is operating. A restriction anywhere in the high side will result in a cold spot or the appearance of frost.

Answer A is incorrect. A faulty compressor discharge valve would not cause this problem. A faulty compressor discharge valve will cause poor cooling accompanied with unusual gauge pressures.

Answer C is incorrect. A clogged expansion valve will not cause a band of frost to appear on the high-pressure hose. If the expansion valve gets restricted, then the line downstream might be frosted.

Answer D is incorrect. Moisture in the system could cause a restriction at the expansion valve and result in frost appearing at the expansion valve.

Question #21
Answer B is correct. Refrigerants are heavier than air and will be most effectively detected below a suspected leak.

Answer A is incorrect. Refrigerants are heavier than air and will not be effectively detected above the suspected leak.

Answer C is incorrect. Six inches upstream from the leak would cause small leaks to go undetected.

Answer D is incorrect. Six inches downstream from the leak would cause small leaks to go undetected.

Question #22
Answer C is correct. Both Technicians A and B are correct. The problem could be either the wiring or the switch. Only visual confirmation or verification with testing equipment can determine the actual fault.

Answer A is incorrect. Technician B is also correct.

Answer B is incorrect. Technician A is also correct.

Answer D is incorrect. Both technicians are correct.

Question #23
Answer A is correct. A plugged air filter will block airflow.

Answer B is incorrect. Filters are critical to airflow.

Answer C is incorrect. Only Technician A is correct.

Answer D is incorrect. Only Technician A is correct.

Question #24
Answer C is correct. Blower operation is delayed to prevent air from entering the bus at an uncomfortable temperature.

Answer A is incorrect. This temperature is too low.

Answer B is incorrect. This temperature is too low.

Answer D is incorrect. This temperature is too high.

Question #25
Answer A is correct. A digital multimeter (DMM) is the most precise tool for electrical/ electronic components or systems and because of its 10 megohm impedance; it will not harm solid-state components.

Answer B is incorrect. A self-powered test lamp could damage solid-state components. The self-powered test light (also called a continuity tester) is only used to check switches and wiring for continuity.

Answer C is incorrect. An analog VOM can damage solid-state components because it draws too much current.

Answer D is incorrect. A 12 V test lamp can damage solid-state components because it draws too much current.

Question #26
Answer B is correct. The reason for the tripped circuit breaker should be repaired.

Answer A is incorrect. The breaker will trip again if the problem is not corrected.

Answer C is incorrect. Only Technician B is correct.

Answer D is incorrect. Only Technician B is correct.

Question #27
Answer A is correct. The first thing to determine is the current flow to the motor.

Answer B is incorrect. The switch is not the first item to check. The left motor is operating.

Answer C is incorrect The resistor would only affect one speed. The motor is not operating.

Answer D is incorrect. The blower motor may not be the problem.

Question #28
Answer A is correct. The Clean Air Act (CAA) established the rule of certification. This rule states that all persons authorized to operate the equipment must be certified under the Act.

Answer B is incorrect. This would be in violation of the CAA.

Answer C is incorrect. This would be in violation of the CAA.

Answer D is incorrect. This would be in violation of the CAA.

Question #29
Answer A is correct. The damper is usually controlled by a cable. This damper directs the airflow toward the defroster grill or slots to blow air onto the windshield.

Answer B is incorrect. The coolant valve is used to control the flow of coolant to the heater core.

Answer C is incorrect. Only Technician A is correct.

Answer D is incorrect. Only Technician A is correct.

Question #30
Answer A is correct. Only Technician A is correct. Any leak from the system must be repaired as quickly as possible.

Answer B is incorrect. If oil is leaking from the system, refrigerant must also be leaking from the system.

Answer C is incorrect. Only Technician A is correct.

Answer D is incorrect. Only Technician A is correct.

Question #31
Answer A is correct. POE oil is specifically formulated to be compatible with R-134a. POE oil is a synthetic-based lubricant that is light enough to be circulated by 134a systems.

Answer B is incorrect. R-22 systems use mineral- or synthetic-based lubricant.

Answer C is incorrect. DEXRON is not compatible with R-134a and will damage A/C system components.

Answer D is incorrect. SAE 30 will damage A/C system components.

Question #32
Answer B is correct. Only Technician B is correct. The high discharge pressure is the result of excessive heat in the system. When the condenser section is inefficient the higher heat results in higher pressure.

Answer A is incorrect. The problem needs to be found and corrected.

Answer C is incorrect. Only Technician B is correct.

Answer D is incorrect. Only Technician B is correct.

Question #33
Answer A is correct. Only Technician A is correct. A fan operating at maximum rpm has a defective fan control unit.

Answer B is incorrect. Only Technician A is correct.

Answer C is incorrect. Only Technician A is correct

Answer D is incorrect. Only Technician A is correct.

Question #34
Answer C is correct. When you test a heater core for leaks, you apply 10 psi of air pressure; a pressure leakage rate of 5 psi in three minute is not acceptable. The heater core would need to be replaced.

Answer A is incorrect. Any pressure bleed-down indicates that the heater core is leaking and must be replaced.

Answer B is incorrect. If the heater core leaks, it should be replaced immediately to prevent coolant loss and the risk of engine damage.

Answer D is incorrect. This is an acceptable method of checking a heater core for leaks.

Question #35
Answer C is correct. There is no visual way to observe circulation through radiator.

Answer A is incorrect. When the thermostat opens, the upper radiator hose rapidly gets warm.

Answer B is incorrect. When the thermostat opens, the temperature gauge rises until it indicates normal operating temperature.

Answer D is incorrect. The computer read-out of the engine temp sensor is a valid indicator of thermostat operation.

Question #36
Answer A is correct. A stuck coolant valve will make HVAC output temperature uncontrollable. When the control is set to cold, the stuck coolant valve allows hot water to flow into the heater core, warming the air. When the control is set to hot, the coolant valve will not shut off the hot coolant to maintain temperature set point. When the control is set to vent, the air is warmed by the heater core.

Answer B is incorrect. A defective compressor clutch will not affect heater temperature control.

Answer C is incorrect. A clogged expansion valve will not affect heater temperature control.

Answer D is incorrect. An inoperative blower motor has no effect on temperature control.

Question #37
Answer D is correct. Under these conditions, a cold windshield causes moisture to condense from the outside air.

Answer A is incorrect. A clogged evaporator drain will cause fogging on the inside of the windshield, accompanied by a mildew smell.

Answer B is incorrect. A leaking heater core will cause fogging on the inside of the windshield, accompanied by a sweet smell.

Answer C is incorrect. An iced-up evaporator core will not cause the windshield to fog. This problem would result in poor A/C cooling.

Question #38
Answer D is correct. Whenever a vehicle is retrofitted from R-12 to R-134a, a blue retrofit decal is to be installed in a visible location. The technician needs to list the charge amount, the amount and type of oil used, and the name of the service business that performed the retrofit.

Answer A is incorrect. Having a new compressor does not guarantee that a vehicle has been retrofitted.

Answer B is incorrect. Although it is advisable to replace the dryer when performing a retrofit, seeing a new one on a vehicle does not mean that it has had a retrofit.

Answer C is incorrect. Seeing the caps missing from the service ports does not prove that the vehicle has been retrofitted.

Question #39
Answer D is correct. Neither technician is correct. The condenser fan may still be operating but a dirty coil could be the problem. The motor drive failure would result in a different code.

Answer A is incorrect. The fan failure is not the only possible problem.

Answer B is incorrect. The driver failure would be code 13.

Answer C is incorrect. Neither technician is correct.

Question #40
Answer D is correct. Synthetic refrigeration oil is the recommended lubricant for A/C O-rings.

Answer A is incorrect. Petroleum jelly must not be used to lubricate O-rings.

Answer B is incorrect. Transmission fluid can damage A/C O-rings.

Answer C is incorrect. Silicone grease must not be introduced into the system.

Question #41
Answer A is correct. The N.O. relay contact EFSH will be closed.

Answer B is incorrect. The EFSL is for slow speed.

Answer C is incorrect. This is a condenser relay.

Answer D is incorrect. This is a condenser relay.

Question #42
Answer A is correct. Only Technician A is correct. Some HVAC control panels contain independent relays such as the blower speed and the mode relays.

Answer B is incorrect. Some bus control panel components are serviceable.

Answer C is incorrect. Only Technician A is correct.

Answer D is incorrect. Only Technician A is correct.

Question #43
Answer A is correct. When you increase the cooling system pressure, you increase the boiling point, not decrease it.

Answer B is incorrect. When more antifreeze is added to the coolant mix, the boiling point is increased.

Answer C is incorrect. High quality ethylene glycol antifreeze contains a corrosion inhibitor.

Answer D is incorrect. Coolant solutions must be recovered, recycled, or handled as hazardous material.

Question #44
Answer B is correct. Raising the pressure compresses the rate of gas expansion, thereby increasing the temperature to reach the boiling point. Each 1 psi of pressure added to the cooling system results in the boiling point of the coolant being raised by 3° F.

Answer A is incorrect. Raising the pressure of the cooling system raises the boiling point of the coolant.

Answer C is incorrect. Raising the pressure in the cooling system does not prevent corrosion. Antifreeze contains corrosion inhibitors that help prevent corrosion.

Answer D is incorrect. Raising the pressure raises the boiling point 3 degrees for each psi of pressure that is added.

Question #45
Answer A is correct. Only Technician A is correct. As the set point is reached, the heat load is being reduced in the bus. As the suction pressure drops, the compressor should cut back its capacity with the unloader system. This prevents the excessive cycling of the clutch.

Answer B is incorrect. If a unit is low on refrigerant, the chances of reaching set point would be almost impossible.

Answer C is incorrect. Only Technician A is correct.

Answer D is incorrect. Only Technician A is correct.

Question #46

Answer C is correct. A digital multimeter is the correct instrument to use to diagnose electronic circuit component malfunctions. One note of caution is to make sure the multimeter is a high impedance device. For safe use on electronic circuits, the meter needs to have at least 10 megohms of internal resistance (impedance).

Answer A is incorrect. A 12-volt test light can damage electronic circuit components. A test light will pull too much current to be used on electronic circuits.

Answer B is incorrect. An analog multimeter can damage electronic circuit components. If a multimeter is required, it must be a DMM (digital multimeter) with high impedance.

Answer D is incorrect. An A/C charging station cannot be used to diagnose electronic circuit component malfunctions.

Question #47

Answer C is correct. Both Technicians A and B are correct. Hand-held diagnostic tools are often used to calibrate and diagnose HVAC systems. However, most systems have a self-diagnostic feature that allows a technician to retrieve codes and even recalibrate the sensors just by pressing the correct buttons on the control head.

Answer A is incorrect. Technician B is also correct.

Answer B is incorrect. Technician A is also correct.

Answer D is incorrect. Both technicians are correct.

Question #48

Answer D is correct. With a coolant valve stuck open, the duct temperature will be constant.

Answer A is incorrect. A low refrigerant charge will not affect heater temperature control.

Answer B is incorrect. Even with a very low coolant level, cooling will take place with the A/C on.

Answer C is incorrect. A compressor clutch failure will not affect heater temperature control.

Question #49

Answer B is correct. Only Technician B is correct. The hose should be replaced immediately. All of the coolant hoses should be thoroughly checked at this time. If the hoses are the same age as the failed hose, it is advisable to replace all of them at the same time.

Answer A is incorrect. A bulge in the hose indicates a weak spot.

Answer C is incorrect. Only Technician B is correct.

Answer D is incorrect. Only Technician B is correct.

Question #50

Answer C is correct. The low-pressure cutout switch (also called the pressure cycling switch) is mounted in the low side of the AC system and opens when the pressure drops below 20 to 25 psi.

Answer A is incorrect. The low-pressure cutout switch is located in the low side of the system.

Answer B is incorrect. An aneroid or BAP sensor senses atmospheric pressure and is part of the fuel delivery system, not the A/C system.

Answer D is incorrect. There is no pressure sensor in the bus interior.

Question #51
Answer A is correct. Only Technician A is correct. The control panel should be replaced when one segment of the digital readout on an ATC control panel is inoperative. Most of the ATC control panels allow no parts to be replaced, except illumination bulbs.

Answer B is incorrect. LEDs on control panels cannot be replaced in the field.

Answer C is incorrect. Only Technician A is correct.

Answer D is incorrect. Only Technician A is correct.

Question #52
Answer D is correct. The addition of refrigerant oil is required when replacing the compressor.

Answer A is incorrect. Oil is not needed in the evaporator prior to installation.

Answer B is incorrect. Oil is not needed in the condenser prior to installation.

Answer C is incorrect. Oil is not needed in the accumulator prior to installation.

Question #53
Answer D is correct. The expansion valve is an adjustable device.

Answer A is incorrect. An expansion valve can be adjusted.

Answer B is incorrect. An expansion valve can be adjusted.

Answer C is incorrect. An expansion valve can be adjusted.

Question #54
Answer A is correct. Only Technician A is correct. A zener diode is used as an input to the computer.

Answer B is incorrect. Sensors are usually supplied with V-Ref (±5 VDC) and not V-Bat.

Answer C is incorrect. Only Technician A is correct.

Answer D is incorrect. Only Technician A is correct.

Question #55
Answer A is correct. Disinfectant will kill the mold and mildew that is the source of the odor. Another method to correct a mildew odor is to use a fungicide spray on the evaporator core.

Answer B is incorrect. This method will simply mask the odor and not eliminate the problem.

Answer C is incorrect. This method poses a possible fire hazard.

Answer D is incorrect. This method will simply mask the odor and not eliminate the problem.

Question #56
Answer C is correct. Both Technicians A and B are correct. R-22 is distributed in green container and R-134a is distributed in blue containers. The service fittings on R-134a containers are larger than fittings on R-22 containers to help prevent cross-contamination.

Answer A is incorrect. Technician B is also correct.

Answer B is incorrect. Technician A is also correct.

Answer D is incorrect. Both technicians are correct.

Question #57
Answer B is correct. Relative humidity will not significantly affect airflow through the condenser. However, the humidity level greatly affects the pressures in the high side of the system. High humidity levels cause the pressure in the high side of the system to elevate dramatically.

Answer A is incorrect. Debris trapped in the condenser fins will significantly affect airflow through the condenser.

Answer C is incorrect. Bent fins will affect airflow through the condenser.

Answer D is incorrect. Vehicle speed will affect airflow through the condenser.

Question #58
Answer C is correct. Both Technicians A and B are correct. The containers are color-coded and the fittings may be different to avoid mixing refrigerants.

Answer A is incorrect. Technician B is also correct.

Answer B is incorrect, Technician A is also correct.

Answer D is incorrect. Both technicians are correct.

Question #59
Answer C is correct. When the system is completely empty, the technician connects the manifold gauge center hose to a vacuum pump. The center hose is typically the yellow hose in a manifold gauge set.

Answer A is incorrect. The high-pressure hose is connected to the high-side service valve.

Answer B is incorrect. The low-pressure hose is connected to the low-side service valve.

Answer D is incorrect. Connecting the wrong hose to the vacuum pump can result in an inadequate evacuation.

Question #60
Answer D is correct. A magneto is used to generate the ignition spark in many small gasoline engines.

Answer A is incorrect. Chassis air pressure can be used to operate a coolant control valve.

Answer B is incorrect. Vacuum can be used to operate a coolant control valve.

Answer C is incorrect. Older systems used a cable to operate the coolant control valve.

Question #61
Answer A is correct. Only Technician A is correct. If the air gap is too great, the clutch will slip. The compressor clutch air gap can be checked by sliding a feeler gauge between the drive and driven plates. However, if the armature and the pulley face are worn with grooves, the feeler gauge will be inaccurate. If the air gap is above specs, clutch adjustment is necessary.

Answer B is incorrect. The drive and driven plates must be replaced, not resurfaced.

Answer C is incorrect. Only Technician A is correct.

Answer D is incorrect. Only Technician A is correct.

Question #62
Answer C is correct. Antifreeze protection should be measured with a refractometer. A hydrometer can also be used to check the antifreeze protection.

Answer A is incorrect. A litmus strip is used to measure relative acidy/alkalinity.

Answer B is incorrect. A manometer is used to measure inlet restriction.

Answer D is incorrect. An opacity meter is used to measure smoke density.

Question #63
Answer D is correct. Neither technician is correct. The growling or rumbling noise would have to be coming from something in motion with the AC turned on or off. Items that could possibly cause this problem would include the water pump, an idler bearing, an alternator bearing, or possibly a loose mounting bracket in the drive belt system. Any one of these could be the problem.

Answer A is incorrect. Technician A is incorrect. If the compressor bearing was the cause, the noise would only be present with the clutch engaged.

Answer B is incorrect. Technician B is incorrect. A defective clutch bearing is not the only source of possible noise.

Answer C is incorrect. Neither technician is correct.

Question #64
Answer C is correct. Both Technicians A and B are correct. The low-pressure cutoff switch prevents compressor operation if the refrigerant charge has been lost or ambient temperature too cold. The high-pressure cutoff switch protects the system from excessive pressure.

Answer A is incorrect. Technician B is also correct.

Answer B is incorrect. Technician A is also correct.

Answer D is incorrect. Both technicians are correct.

Question #65
Answer D is correct. Electric cooling fan motors are never controlled by an air solenoid controller.

Answer A is incorrect. Electric cooling fans are often controlled by an electronic relay.

Answer B is incorrect. A cooling fan module can control electric cooling fans.

Answer C is incorrect. A multiplexed electronic module can control electric cooling fans.

Question #66
Answer A is correct. If the booster pump circuit breaker is open, the compressor clutch would not be disabled.

Answer B is incorrect. An ambient temperature below 45° F will disable the system.

Answer C is incorrect. Ambient air temperature below 40° F will open the ambient cut-off switch, disabling the compressor clutch.

Answer D is incorrect. An open circuit in 31B will disable the compressor clutch.

Question #67
Answer B is correct. The laptop will read out the code and offer system troubleshooting advice.

Answer A is incorrect. The code will not be eliminated and the problem will still exist.

Answer C is incorrect. The routine maintenance is important, but in this case the problem needs to be addressed.

Answer D is incorrect. The filter and nozzle change are considered routine maintenance.

Answers to the Test Questions for the Additional Test Questions Section 6

1. B
2. A
3. B
4. D
5. C
6. D
7. B
8. B
9. C
10. C
11. B
12. A
13. A
14. C
15. A
16. A
17. A
18. D
19. B
20. D
21. A
22. D
23. C
24. D
25. C
26. A
27. A
28. B
29. B
30. A
31. C
32. A
33. C
34. D
35. A
36. D
37. B
38. C
39. D
40. B
41. A
42. C
43. D
44. A
45. B
46. A
47. B
48. C
49. B
50. A
51. C
52. C
53. A
54. C
55. A
56. B
57. A
58. C
59. A
60. A
61. A
62. D
63. C
64. B
65. D
66. C
67. B
68. A
69. A
70. C
71. C
72. C
73. A
74. D
75. D
76. B
77. D
78. A
79. B
80. C
81. B
82. C
83. D
84. C
85. D
86. D
87. C
88. D
89. B
90. A
91. B
92. B
93. B
94. A
95. D
96. A
97. D
98. B
99. C
100. A
101. A
102. C
103. A
104. B
105. D
106. A
107. A
108. B
109. D
110. B
111. D
112. A
113. B
114. B
115. C
116. D
117. B

Explanations to the Answers for the Additional Test Questions Section 6

Question #1
Answer B is correct. Only Technician B is correct. A small thermometer can be used to monitor the performance of an HVAC system. The thermometer should be placed in the center duct during a performance test to show the temperature of the discharge air.

Answer A is incorrect. Many other tools are required to properly diagnose HVAC system failures.

Answer C is incorrect. Only Technician B is correct.

Answer D is incorrect. Only Technician B is correct.

Question #2
Answer A is correct. The engine radiator shutters are usually controlled mechanically or by the engine control module.

Answer B is incorrect. The HVAC system controls the speed of the blower motor.

Answer C is incorrect. The HVAC system controls the fresh air damper actuator to modulate the output air temperature.

Answer D is incorrect. The HVAC system controls the position of the outside air door.

Question #3
Answer B is correct. Air is not condensable at A/C system pressures. To identify if a vessel of refrigerant has excessive air, a technician can check the pressure in the vessel and compare this pressure with the pressure/temperature chart.

Answer A is incorrect. Any acids that form in the A/C system are created due to moisture and corrosion and are usually in a liquid state.

Answer C is incorrect. Moisture in the A/C system only evaporates when the system is drawn under a vacuum, and is therefore normally in a liquid state.

Answer D is incorrect. Refrigeration oil is normally in a liquid state.

Question #4
Answer D is correct. It is acceptable to vent nitrogen to the atmosphere. The atmosphere already contains an abundant amount of nitrogen, so venting nitrogen during A/C service will do no harm.

Answer A is incorrect. To prevent damaging A/C system components, the nitrogen pressure must be regulated.

Answer B is incorrect. The A/C compressor must be disconnected before flushing the system to prevent debris from entering into it.

Answer C is incorrect. To avoid damaging restrictive components, they must be removed before flushing.

Question #5
Answer C is correct. Even a small leak from the weep hole indicates that the front seal of the water pump has failed.

Answer A is incorrect. Fan clutch replacement does not affect the life of the water pump.

Answer B is incorrect. The water pump generally has a longer service life than the heater hoses.

Answer D is incorrect. Thermostat operation does not indicate a need to replace the water pump.

Question #6
Answer D is correct. OEMs recommend either a high-side (liquid) or a low-side (vapor) charging process.

Answer A is incorrect. A technician completes the charging process when the correct weight of refrigerant has entered the system.

Answer B is incorrect. If the low side does not move from a vacuum to a pressure, there is a restriction.

Answer C is incorrect. The bus engine does not need to be running when recharging with a recharging machine.

Question #7
Answer B is correct. R-134a has a faint ether-like odor. It is possible for a driver to comment that they have an ether-like smell in their vehicle if the evaporator core has a large leak.

Answer A is incorrect. R-134a does have an ether-like odor.

Answer C is incorrect. R-134a does not have a strong rotten egg odor.

Answer D is incorrect. R-134a does not have a cabbage-like odor.

Question #8
Answer B is correct. The coolant control valve is not part of the water pump. It is usually located somewhere in the inlet heater hose.

Answer A is incorrect. The coolant control valve does control the flow of coolant through the heater core.

Answer C is incorrect. The coolant control valve may be cable operated.

Answer D is incorrect. The coolant control valve may be electrically operated.

Question #9
Answer C is correct. A flame-type leak detector is not recommended. Using a flame-type leak detector with R-22 produces toxic phosgene gas. Flame-type leak detectors do not change colors when exposed to R-134a, so they are not useful.

Answer A is incorrect. This is an effective method for finding large leaks.

Answer B is incorrect. An electronic leak detector is one of the most accurate methods of leak detection.

Answer D is incorrect. Using a black light detector is recommended and effective.

Question #10
Answer C is correct. Both Technicians A and B are correct. The resistance of the clutch coil can be tested with an ohmmeter. Another test that can be performed is supplying power and ground to the coil to see if the clutch will engage.

Answer A is incorrect. Technician B is also correct.

Answer B is incorrect. Technician A is also correct.

Answer D is incorrect. Both technicians are correct.

Question #11
Answer B is correct. 10–45 psi is the normal range of low-side pressures.

Answer A is incorrect. 5–10 psi is an extremely low, low-side pressure indicative of a clogged orifice tube or low refrigerant charge.

Answer C is incorrect. 60–80 psi is a very high, low-side pressure.

Answer D is incorrect. 180–205 psi is a typical high-side pressure.

Question #12
Answer A is correct. Only Technician A is correct. A cloudy sight glass indicates that the desiccant pack in the filter/dryer has broken.

Answer B is incorrect. The filter/dryer must be replaced if the desiccant pack breaks down or if the system has been open for an extended period of time.

Answer C is incorrect. Only Technician A is correct.

Answer D is incorrect. Only Technician A is correct.

Question #13
Answer A is correct. Restricted airflow through the condenser will cause elevated high-side pressure.

Answer B is incorrect. A thermostat that is stuck open will cause poor heater performance.

Answer C is incorrect. The thermal bulb will not leak.

Answer D is incorrect. A restriction in the filter dryer will cause low, high-side pressures.

Question #14
Answer C is correct. The thermal bulb and capillary tube are part of the thermal expansion valve located near the evaporator inside of the duct housing. The thermal bulb and capillary tube sense the temperature of the outlet line of the evaporator core. When the thermal bulb senses warm temperatures, the TXV opens to allow more refrigerant into the evaporator core. When the thermal bulb senses cold temperatures, the TXV closes to restrict refrigerant flow into the evaporator.

Answer A is incorrect. The expansion valve is located near the evaporator, not the condenser.

Answer B is incorrect. An accumulator is used with a CCOT system that does use a thermal bulb or capillary tube.

Answer D is incorrect. A CCOT system does use a thermal bulb or capillary tube.

Question #15
Answer A is correct. Comparing the pressure of recovered refrigerant to the theoretical pressure of pure refrigerant at a given temperature is the best method of testing for non-condensable gases in refrigerant.

Answer B is incorrect. Pressure cannot be compared to humidity.

Answer C is incorrect. Pressure cannot be compared to volume.

Answer D is incorrect. A halogen leak detector cannot be used to check for non-condensable gases.

Question #16
Answer A is correct. The valve is front-seated when the stem is turned clockwise to seat the front valve face to the left. This shuts off the flow of refrigerant to the compressor and isolates it.

Answer B is incorrect. Back-seating the valve moves it counter clockwise to the right to seal the rear face valve. This is the normal valve position.

Answer C is incorrect. When the valve is at mid-position, it is not seated at all. The valve should be in this position only after connecting the hose for the manifold gauge set. It remains in this position while conducting any service to the system, but must be back-seated before removing the service hose.

Answer D is incorrect. In normal operating position, the valve is back-seated.

Question #17
Answer A is correct. Only technician A is correct. The high-pressure refrigerant at the compressor outlet can raise the temperature of the outlet to that of the engine coolant or even higher on a hot and humid day.

Answer B is incorrect. This is a normal condition caused by the compressor raising the pressure/temperature of the vapor refrigerant.

Answer C is incorrect. Only Technician A is correct.

Answer D is incorrect. Only Technician A is correct.

Question #18
Answer D is correct. A blockage of the expansion valve will cause low, high-side pressure.

Answer A is incorrect. A refrigerant overcharge will cause high system pressures.

Answer B is incorrect. A dirty condenser coil will cause condenser pressure and high condenser temperature.

Answer C is incorrect. A restricted airflow through the condenser will cause elevated high-side pressure due to the lack of heat transfer resulting from restricted airflow.

Question #19
Answer B is correct. The A/C high-pressure switch opens the electrical circuit to the compressor clutch coil when high-side pressure reaches its upper limit. The high-pressure switch typically opens at about 380–420 psi.

Answer A is incorrect. The high-pressure switch does not provide a boosting function.

Answer C is incorrect. The high-pressure switch does not maintain pressure.

Answer D is incorrect. The high-pressure switch does perform a venting function.

Question #20
Answer D is correct. If this circuit breaker opens, it affects only the panel light and will not prevent clutch engagement.

Answer A is incorrect. A defective blower switch could prevent clutch engagement by not allowing current to flow from the white wire to the mode control switch.

Answer B is incorrect. An open binary pressure control switch will prevent clutch engagement because it is in the path to the A/C clutch.

Answer C is incorrect. A poor connection at the thermostat switch could prevent clutch engagement.

Question #21
Answer A is correct. Cooling and heating system hoses do not need to be replaced because they leak at the hose clamps; usually tightening the clamps repairs the leak.

Answer B is incorrect. Cracked hoses must be replaced.

Answer C is incorrect. Bulging hoses must be replaced.

Answer D is incorrect. Spongy hoses must be replaced.

Question #22
Answer D is correct. Neither technician is correct. It is normal to have an audible noise in the AC system after shutting the engine off. The noise comes from the equalizing pressures from each side of the system.

Answer A is incorrect. Technician A is incorrect. A faint hissing noise after shutting down the engine is caused by the equalization of refrigerant pressure in the A/C system.

Answer B is incorrect. Technician B is incorrect. A defective TXV would cause poor cooling accompanied by irregular pressures.

Answer C is incorrect. Neither technician is correct.

Question #23
Answer C is correct. Both Technicians A and B are correct. If moisture enters the A/C system, it mixes with the refrigerant to form harmful acids. The method of removing any moisture from the A/C system is to pull a vacuum on the system for at least 30 minutes. Pulling a vacuum on the system will cause any moisture to vaporize and be pulled out of the system.

Answer A is incorrect. Technician B is also correct.

Answer B is incorrect. Technician A is also correct.

Answer D is incorrect. Both technicians are correct.

Question #24
Answer D is correct. Neither technician is correct. R-22 systems use a mineral- or synthetic-based lubricant; R-134a systems use POE oil, a synthetic lubricant.

Answer A is incorrect. Technician A is incorrect. Only R-22 A/C systems can use mineral oil.

Answer B is incorrect. Technician B is incorrect. POE oil is a synthetic lubricant used in R-134a A/C systems.

Answer C is incorrect. Neither technician is correct.

Question #25
Answer C is correct. Foam tape can be used to cushion and seal around the heater core. A technician should be very careful when replacing a heater core to make sure that the new core is secure and sealed.

Answer A is incorrect. The heater core does not need insulation.

Answer B is incorrect. The heater core does not contribute significantly to cab noise.

Answer D is incorrect. A gasket is provided to seal around heater hose connections.

Question #26
Answer A is correct. A compressor with worn reed valves would be unable to create very much pressure difference between high- and low-side readings.

Answer B is incorrect. No pressure would most likely mean the system is empty.

Answer C is incorrect. Both gauges reading too low would likely indicate an undercharge, or a restriction.

Answer D is incorrect. When both gauges read too high, the cause could be an overcharge, high operating temperatures, air or moisture in the system, or a TXV stuck open.

Question #27
Answer A is correct. The compressor discharge valve opens after the pressure compresses vaporized refrigerant, allowing the refrigerant to move to the condenser to change from a high-pressure gas into a high-pressure liquid.

Answer B is incorrect. If the discharge valve opened before vaporized refrigerant is compressed, it would not properly raise the refrigerant pressure to allow a change from gas to a liquid while in the condenser. In addition, it does not go from the compressor to the evaporator.

Answer C is incorrect. The reed valves just open and close, they do not regulate a variable pressure.

Answer D is incorrect. Reed valves cannot sense A/C system temperature, so regulation cannot take place.

Question #28
Answer B is correct. Only Technician B is correct. Moisture that enters the A/C system will be harmful to the system and cause poor performance because of internal erosion and the possibility of the moisture turning to ice near the expansion device.

Answer A is incorrect. Moisture cannot be removed from the A/C system after charging the system with new refrigerant. However, all systems have a drying device that should capture any trace amounts of moisture that enters the system.

Answer C is incorrect. Only Technician B is correct.

Answer D is incorrect. Only Technician B is correct.

Question #29
Answer B is correct. Only Technician B is correct. If liquid refrigerant enters the compressor, then the compressor will be damaged because liquid cannot be compressed.

Answer A is incorrect. If the technician charges into the high-side with the engine running, high-pressure refrigerant may be forced into the container or charging machine.

Answer C is incorrect. Only Technician B is correct.

Answer D is incorrect. Only Technician B is correct.

Question #30
Answer A is correct. If the expansion valve is stuck closed, the low side of the system will be pulled down into a vacuum and the compressor will not be able to sufficiently pressurize the high side.

Answer B is incorrect. Without flow from the evaporator, the compressor cannot generate high-pressure.

Answer C is incorrect. With the low side being restricted, the low-side pressure will be nearing a vacuum.

Answer D is incorrect. This condition would be caused by a faulty compressor.

Question #31
Answer C is correct. Both Technicians A and B are correct. A malfunctioning TXV could cause the evaporator core to ice up, causing reduced airflow to the vent outlets.

Answer A is incorrect. Technician B is also correct.

Answer B is incorrect. Technician A is also correct.

Answer D is incorrect. Both technicians are correct.

Question #32
Answer A is correct. Only Technician A is correct. The best way to repair wiring is by soldering. Heat shrink wrap should also be used to keep out moisture and prevent future corrosion problems. One other quality wire repair technique is using crimp and seal connectors. These connectors also keep out moisture and prevent future corrosion problems.

Answer B is incorrect. Twisting wires together and taping them will result in a high resistance connection and ultimately circuit failure.

Answer C is incorrect. Only Technician A is correct.

Answer D is incorrect. Only Technician A is correct.

Question #33
Answer C is correct. Both Technicians A and B are correct. A low refrigerant charge will cause this symptom. The low-pressure cutout switch opens its contacts when the pressure drops below 10 psi. The switch acts as a de-icing device when the system is operating. The switch also serves as a low refrigerant protection device that keeps the compressor from turning on when the refrigerant gets low.

Answer A is incorrect. Technician B is also correct..

Answer B is incorrect. Technician A is also correct.

Answer D is incorrect. Both technicians are correct

Question #34
Answer D is correct. Return air filters are designed to remove dust and dirt from bus evaporator inlet air. These filters should be inspected while performing a preventative maintenance inspection and should be replaced if found to be dirty or restricted.

Answer A is incorrect. Some bus air filters are made of paper or a metallic mesh.

Answer B is incorrect. Bus air filters are not designed to remove moisture from the air.

Answer C is incorrect. Bus air filters are replaceable and should be checked during routine maintenance checks.

Question #35
Answer A is correct. This symptom is typical of an A/C system with an evaporator icing problem, which could be caused by a defective thermal expansion valve.

Answer B is incorrect. A defective coolant control valve would not cause the A/C system output air to change from cold to warm. Usually coolant valves stick open, causing warm air all the time.

Answer C is incorrect. The coolant pump is not used during high cool.

Answer D is incorrect. The fresh air damper does not control the output air temperature.

Question #36
Answer D is correct. Neither technician is correct. Both technicians are stating that refrigerant oil needs to be added during heater core replacement, and this is not required.

Answer A is incorrect. Technician A is incorrect. Oil is not added to the heater core prior to installation.

Answer B is incorrect. Technician B is incorrect. Oil is not added to the heater core prior to installation.

Answer C is incorrect. Neither technician is correct.

Question #37
Answer B is correct. The heater hoses connected to the coolant control valve may be clamped during the replacement procedure to maintain coolant in the system. Many tool companies sell hose-pinching pliers, a tool to perform the clamping function.

Answer A is incorrect. The refrigerant does not need to be recovered from the system before the valve is replaced.

Answer C is incorrect. The coolant control valves can be located under the bus or in the HVAC equipment.

Answer D is incorrect. The coolant control valve is an individually replaceable component.

Question #38
Answer C is correct. The water temperature sensor only controls blower operation in the heat mode.

Answer B is incorrect. The ATS sensor signal is used to control compressor operation.

Answer A is incorrect. The compressor temperature sensor signal is used to control compressor operation.

Answer D is incorrect. The compressor discharge sensor signal is used to control compressor operation.

Question #39
Answer D is correct. Nothing should be added to ELC premix since it is already at the correct mixture level.

Answer A is incorrect. Coolant conditioner should never be added to ELC.

Answer B is incorrect. Premixed ELC is already at the correct strength and water should never be added to it.

Answer C is incorrect. Never add propylene glycol coolant to a system that has ELC.

Question #40
Answer B is correct. Only Technician B is correct. New environmental laws dictate that shut-off valves must be located no more than 12 inches from test hose service end.

Answer A is incorrect. Shut-off valves do not have to be closed every time the A/C system is switched off. However, they must be closed when connecting and disconnecting the hoses from the vehicle.

Answer C is incorrect. Only Technician B is correct.

Answer D is incorrect. Only Technician B is correct.

Question #41
Answer A is correct. A 50/50 mix will result in a very slow windshield defrost.

Answer B is incorrect. Directing 85 percent of the output air to the windshield in the defrost mode is an appropriate mix.

Answer C is incorrect. A 75 percent airflow to the windshield will defrost the windshield better than a 50 percent airflow to the windshield.

Answer D is incorrect. A 70 percent airflow to the windshield will defrost the windshield better than a 50 percent airflow to the windshield.

Question #42
Answer C is correct. The technician needs to drain the entire coolant system and add the proper SCA mixture to achieve the correct percentage of SCA.

Answer A is incorrect. SCA is a corrosion inhibitor additive. Adding more antifreeze does not change the SCA reading.

Answer B is incorrect. Continuing to run the truck until the next PMI will not fix the problem. It is possible that if the coolant was only slightly over-conditioned, the truck could be run until the next service interval and retested. However, running a diesel engine with a severely over-conditioned cooling system can cause engine damage.

Answer D is incorrect. Action should be taken to correct the problem as soon as possible.

Question #43
Answer D is correct. The circuit schematic should be consulted to determine if a resistor is used in the speed control of the motor.

Answer A is incorrect. The circuit diagram should be checked first.

Answer B is incorrect. The circuit diagram should be checked first.

Answer C is incorrect. The circuit diagram should be checked first.

Question #44
Answer A is correct. Only Technician A is correct. The expansion valve screen is installed to prevent particulate from circulating through the system in the event that the desiccant bag in the accumulator breaks down.

Answer B is incorrect. Atomization is not important to the evaporation of refrigerant.

Answer C is incorrect. Only Technician A is correct.

Answer D is incorrect. Only Technician A is correct.

Question #45
Answer B is correct. Refrigerant is heavier than air, so it is most easily detected just below a leaking fitting.

Answer A is incorrect. The sensor tip should be as close as possible to the fitting.

Answer C is incorrect. The sensor probe should be just below the fitting.

Answer D is incorrect. The leak could go undetected using this method. The sensor probe should be just below the fitting.

Question #46
Answer A is correct. Most late-model A/C systems are independently controlled.

Answer B is incorrect. Most late-model A/C systems are independently controlled.

Answer C is incorrect. Only Technician A is correct.

Answer D is incorrect. Technician A is correct.

Question #47
Answer B is correct. When more antifreeze is added to the coolant, the boiling point increases.

Answer A is incorrect. When the cooling system pressure is increased, the boiling point increases. For every 1 psi of pressure that is added to the cooling system, the boiling point is raised 3° F.

Answer C is incorrect. A good quality ethylene glycol antifreeze contains antirust inhibitors.

Answer D is incorrect. Coolant solutions must be recovered, recycled, and handled as hazardous waste.

Question #48
Answer C is correct. The component shown is an evaporator pressure regulator EPR valve. If it was not used, the evaporator temperatures could go below freezing and ice up the evaporator core.

Answer A is incorrect. The component shown is an EPR valve; this will have little effect on high side pressures.

Answer B is incorrect. The component shown is an EPR valve; this has no effect on refrigerant filtration.

Answer D is incorrect. The component shown is an EPR valve; this will have no effect on noise.

Question #49
Answer B is correct. A technician uses a pressure tester to test radiators, pressure caps, and hoses for leaks. Caution should be taken to not apply pressure higher than the system is rated.

Answer A is incorrect. Thermostats can be tested by monitoring engine temperature during the warm-up process to make sure the thermostat does not open too early. The thermostat can also be tested by removing it from the bus and submerging it in hot water to see when it opens.

Answer C is incorrect. A/C leaks must be located using a leak detector.

Answer D is incorrect. Vacuum diaphragms are tested using a hand-held vacuum pump.

Question #50
Answer A is correct. Using a belt tension gauge ensures that the belt is properly adjusted.

Answer B is incorrect. This method is not correct and would be excessively tight if used.

Answer C is incorrect. This method is not correct because the belt needs to have some static deflection. Using a belt tension gauge insures the correct tension.

Answer D is incorrect. This method could cause component damage due to over-tightening.

Question #51
Answer C is correct. Chilling and heating water and then recirculating it are only used in stationary HVAC systems.

Answer A is incorrect. The mechanical thermostat HVAC system modulates temperature control by cycling the clutch or the circulation pump and coolant valve.

Answer B is incorrect. Microprocessor-controlled systems control all aspects of the HVAC system with a single input from the driver.

Answer D is incorrect. ECT systems require the driver to choose the mode and blower speed while the thermostat controls the temperature.

Question #52
Answer C is correct. The service ports on Type 2 systems are standard #4 flare fittings.

Answer A is incorrect. Compression fittings are not used in Type 2 A/C R-134a systems. The use of SAE-approved quick-connect couplings on the service ports is specified by the property ordering the bus.

Answer B is incorrect. The SAE-approved coupling is not required.

Answer D is incorrect. 10-mm threaded ports are not used on R-134a systems.

Question #53
Answer A is correct. This failure code indicates a possible problem with the return air sensor. The technician should follow a diagnostic alarm code chart to troubleshoot the fault.

Answer B is incorrect. Answer A is correct.

Answer C is incorrect. Testing the return air sensor circuit is indicated.

Answer D is incorrect. Testing the return air sensor circuit is indicated.

Question #54
Answer C is correct. Both Technicians A and B are correct. J1991 sets limits of 15 ppm for moisture and 330 ppm for noncondensibles (air).

Answer A is incorrect. Technician B is also correct.

Answer B is incorrect. Technician A is also correct.

Answer D is incorrect. Both technicians are correct.

Question #55
Answer A is correct. A/C fittings should not be disturbed if they are not leaking.

Answer B is incorrect. An A/C maintenance service should include cleaning the condenser fins.

Answer C is incorrect. An A/C maintenance service should include straightening bent condenser fins.

Answer D is incorrect. An A/C maintenance service should include checking all component mounts and insulators.

Question #56
Answer B is correct. The system must be run to stabilize the temperatures and allow the refrigerant in the crankcase to boil off.

Answer A is incorrect. The A/C system does not need to be evacuated for 30 minutes prior to checking the lubricant. The refrigerant does need to be recovered with an approved recovery machine.

Answer C is incorrect. Modern compressors do not have a dipstick to allow the oil to be checked.

Answer D is incorrect. It does not matter if the oil is old; the system needs to be operated before the oil can be checked accurately.

Question #57
Answer A is correct. Only Technician A is correct. Coolant condition should always be tested before replacing a coolant conditioning cartridge to prevent over-conditioning of the coolant.

Answer B is incorrect. Routine changing of coolant conditioning/filter cartridges can result in over-conditioned coolant and failure of components in the cooling system.

Answer C is incorrect. Only Technician A is correct.

Answer D is incorrect. Only Technician A is correct.

Question #58
Answer C is correct. Both Technicians A and B are correct. The oil level needs to be checked and corrected following a loss of a large amount of the system oil. It is also very important to use the type of oil recommended by the manufacturer.

Answer A is incorrect. Technician B is also correct.

Answer B is incorrect. Technician A is also correct.

Answer D is incorrect. Both technicians are correct.

Question #59
Answer A is correct. The thermal bulb must be in contact with the evaporator outlet tube. The thermal bulb serves as the temperature-sensing part of the TXV. If the bulb senses warm temperatures, the TXV opens to allow more refrigerant to enter the evaporator core. If the bulb senses cold temperatures, the TXV closes to restrict the flow of refrigerant into the evaporator core.

Answer B is incorrect. This area is the wrong location. The condenser is in the high side of the system.

Answer C is incorrect. This is close to where the thermal bulb is, but the bulb is located inside the HVAC duct attached to the evaporator outlet line.

Answer D is incorrect. The thermal bulb does not directly contact the refrigerant.

Question #60
Answer A is correct. If the fan clutch is always engaged, it will enhance the operation of the A/C system.

Answer B is incorrect. A stuck coolant valve will cause poor A/C performance.

Answer C is incorrect. A low refrigerant charge will cause poor A/C system performance.

Answer D is incorrect. A refrigerant overcharge will cause poor A/C system performance.

Question #61
Answer A is correct. Only Technician A is correct. A faulty reed valve could cause the high-side pressure to be low and the low-side pressure to be high.

Answer B is incorrect. An overcharge of refrigerant oil will not cause these symptoms but will reduce the cooling capabilities of the system. The likely situation on a system with too much oil is high pressures in both sides of the system.

Answer C is incorrect. Only Technician A is correct.

Answer D is incorrect. Only Technician A is correct.

Question #62
Answer D is correct. A/C system pressures do not vary with cab humidity. However, outside humidity will change A/C system pressures dramatically.

Answer A is incorrect. A/C system pressures will vary with altitude changes.

Answer B is incorrect. A/C system pressures will vary with ambient temperature. Higher ambient temperature causes the pressures to rise, especially the high-side pressure.

Answer C is incorrect. A/C system pressures vary dramatically with ambient humidity changes.

Question #63
Answer C is correct. Both Technicians A and B are correct. A clogged heater core or a misadjusted heater control valve could cause insufficient heater output. Either problem could cause a low amount of heated water to pass through the core.

Answer A is incorrect. Technician B is also correct.

Answer B is incorrect. Technician A is also correct.

Answer D is incorrect. Both technicians are correct.

Question #64
Answer B is correct. Only Technician B is correct. Several bent fins and a moderate accumulation of dead insects will restrict airflow through the condenser enough to cause poor cooling and higher-than-normal system pressures. It is a good practice to check for condenser restrictions while performing a good preventive maintenance inspection.

Answer A is incorrect. The condenser should not be replaced; the fins should be straightened and then it should be cleaned.

Answer C is incorrect. Only Technician B is correct.

Answer D is incorrect. Only Technician B is correct.

Question #65
Answer D is correct. Neither technician is correct. Figure A is an R-12 fitting and figure B is an R-134a fitting.

Answer A is incorrect. Technician A is incorrect. Figure A is an R-12 service port fitting. The low side fitting is 7/16 inch by 20 TPI and the high side is 3/8 inch by 24 TPI.

Answer B is incorrect. Technician B is incorrect. Figure B is an R-134a service port fitting. The low side fitting size is 13 mm and the high side fitting size is 16 mm.

Answer C is incorrect. Neither technician is correct.

Question #66
Answer C is correct. Both Technicians A and B are correct. Systems use an accumulator in the suction line to evaporate any liquid refrigerant. TXV systems use a receiver tank in the liquid line to store the refrigerant.

Answer A is incorrect. Technician B is also correct.

Answer B is incorrect. Technician A is also correct.

Answer D is incorrect. Both technicians are correct.

Question #67
Answer B is correct. To ensure that the entire system is under deep enough vacuum to remove all moisture, the pump should be run until a reading of 1200 microns is achieved. After evacuating, the technician should let the system sit with the vacuum pump turned off and make sure that the vacuum is maintained. If the system does hold a vacuum, then there are no leaks in the system and it can be charged with refrigerant.

Answer A is incorrect. 2500 microns is not a thorough evacuation.

Answer C is incorrect. 500 microns is an excessive evacuation.

Answer D is incorrect. 5000 microns is not a thorough evacuation.

Question #68
Answer A is correct. Technician A is correct. The high-pressure relief valve is considered a one-time valve. When system pressures return to normal, sometimes the valves leak. The high-pressure relief valve blows at pressures of 450–550 psi. Also, the high-pressure relief valve is individually replaceable.

Answer B is incorrect. Technician A is correct.

Answer C is incorrect. Only Technician A is correct.

Answer D is incorrect. Only Technician A is correct.

Question #69
Answer A is correct. Only Technician A is correct. Refrigerant enters the compressor as a low-pressure gas. The compressor raises the pressure and temperature of the refrigerant, and it leaves the compressor as a hot/high-pressure gas through the discharge line.

Answer B is incorrect. Refrigerant changes state from a gas to liquid, that is, it "condenses" in the condenser. Boiling is the change of state from a liquid to a gas.

Answer C is incorrect. Only Technician A is correct.

Answer D is incorrect. Only Technician A is correct.

Question #70
Answer C is correct. Both Technicians A and B are correct. A sticky film on the windshield is an indication of an engine coolant leak at the heater core.

Answer A is incorrect. Technician B is also correct.

Answer B is incorrect. Technician A is also correct.

Answer D is incorrect. Both technicians are correct.

Question #71
Answer C is correct. Both Technicians A and B are correct. When evacuating a system, the vacuum pump must run until 1200 microns to remove moisture from the system. After five minutes of pump operation, the low-side gauge should indicate 20 in. Hg, and the high-side gauge should read below zero (unless the needle is restricted by a stop pin). If the high-side gauge does not drop below zero, there may be a refrigerant blockage. Within 15 minutes, the low-side gauge should drop to about 24 to 26 in. Hg. If it does not, close the low-side gauge valve and observe the gauge. If it slowly rises, there may be a leak.

Answer A is incorrect. Technician B is also correct.

Answer B is incorrect. Technician A is also correct.

Answer D is incorrect. Both technicians are correct.

Question #72
Answer C is correct. Both Technicians A and B are correct. There is a definite problem with this vehicle's temperature system. The return air temperature sensor could be out of calibration and providing a false input to the processor. This problem could also be caused by a sticking fresh air damper.

Answer A is incorrect. Technician B is also correct.

Answer B is incorrect. Technician A is also correct.

Answer D is incorrect. Both technicians are correct.

Question #73
Answer A is correct. Only Technician A is correct. Inline filters are recommended as a method to prevent debris from entering into the compressor.

Answer B is incorrect. Inline oil filters are not a substitute for the inline suction filter.

Answer C is incorrect. Only Technician A is correct.

Answer D is incorrect. Only Technician A is correct.

Question #74
Answer D is correct. When R-22 is burned it creates phosgene gas. Phosgene gas is dangerous to humans.

Answer A is incorrect. R-22 is not flammable.

Answer B is incorrect. The gas that is formed is toxic, attacking the nervous system.

Answer C is incorrect. R-22 does not form chlorine gas in the presence of a flame.

Question #75
Answer D is correct. An improperly adjusted mode door cable could cause the air to not be directed to the desired location.

Answer A is incorrect. The microprocessor does not control the driver heater/defroster.

Answer B is incorrect. A bad blend door actuator would only affect the temperature of the air.

Answer C is incorrect. A blower resistor problem would only affect the blower speeds.

Question #76
Answer B is correct. Heavy-duty bus cooling systems are typically pressurized to about 4 psi.

Answer A is incorrect. A pressure of 1 psi is far too low. For every 1 psi of pressure that is added to the cooling system, the boiling point is raised 3° F.

Answer C is incorrect. A pressure of 20 psi is too high and could possibly cause leaks in cooling system components.

Answer D is incorrect. A pressure of 30 psi is very high and would likely cause leaks in cooling system components.

Question #77
Answer D is correct. A refractometer is the recommended instrument for accurately measuring antifreeze protection in a coolant.

Answer A is incorrect. A hydrometer is commonly used to measure the specific gravity of coolant but does not provide as accurate a measurement of antifreeze protection as a refractometer.

Answer B is incorrect. SCA test strips measure the degree of supplemental coolant additive protection.

Answer C is incorrect. Litmus tests measure the acidity/alkalinity of coolant.

Question #78
Answer A is correct. Only Technician A is correct. An inline filter is sometimes used in the line from the condenser to the evaporator.

Answer B is incorrect. Petroleum jelly is not a recommended lubricant for A/C O-rings. The recommended lubricant for all A/C O-rings is the oil that is used in the system.

Answer C is incorrect. Only Technician A is correct

Answer D is incorrect. Only Technician A is correct.

Question #79
Answer B is correct. The valve will reset itself when A/C system pressure returns to a safe level; however, it may leak after a discharge. The valve is designed to release pressure at 450–550 psi.

Answer A is incorrect. The relief valve cannot be calibrated.

Answer C is incorrect. The relief valve does need to be replaced if it vents refrigerant to the atmosphere and then resets itself.

Answer D is incorrect. All bus A/C systems use a high-pressure relief valve.

Question #80
Answer C is correct. The coolant valve closing may cause this symptom.

Answer A is incorrect. The evaporator motor will operate in any of the modes.

Answer B is incorrect. This would not have an effect when set point is reached.

Answer D is incorrect. Arcing in the evaporator motor will not cause the hammer noise.

Question #81
Answer B is correct. The expansion valve allows low-pressure liquid to be metered into the evaporator. Since the pressure is much lower after the expansion valve, the low-pressure refrigerant starts to boil and take on heat from the surrounding surface and air.

Answer A is incorrect. The expansion valve is located in the evaporator core inlet.

Answer C is incorrect. The expansion valve does not affect the refrigerant flow through the condenser.

Answer D is incorrect. The expansion valve does not affect airflow through the evaporator.

Question #82
Answer C is correct. A manual switch is typically not used to control an electric condenser fan.

Answer A is incorrect. The motor control unit can control an electric condenser fan motor.

Answer B is incorrect. The HVAC ECM can control an electric condenser fan motor.

Answer D is incorrect. An electric condenser fan motor can be controlled by an electronic relay.

Question #83
Answer D is correct. Coolant leaks are the most common maintenance problem in bus HVAC systems. A pressure tester is useful in assisting the technician in finding coolant leaks. When pressurizing the system, never apply more than the cap-rated pressure to prevent unwanted component failure.

Answer A is incorrect. In properly maintained cooling systems, clogged heater cores are rare.

Answer B is incorrect. Control units are relatively reliable.

Answer C is incorrect. Heater control valves do fail, but this is not the most common problem in HVAC systems.

Question #84
Answer C is correct. A good visual inspection will locate most external coolant leaks.

Answer A is incorrect. A black light is only useful when a special dye is introduced into the cooling system.

Answer B is incorrect. A leak detector is used to find refrigerant leaks.

Answer D is incorrect. White smoke in the engine exhaust is evidence of burning antifreeze.

Question #85
Answer D is correct. Acids of combustion are typically found in the engine oil, not in the coolant.

Answer A is incorrect. Flushing the cooling system removes rust from the system.

Answer B is incorrect. Flushing the cooling system removes contaminants from the system.

Answer C is incorrect. Flushing the cooling system can increase the life of the system components.

Question #86
Answer D is correct. Neither technician is correct.

Answer A is incorrect. Technician A is incorrect. There is no need to drain the compressor.

Answer B is incorrect. Technician B is incorrect.

Answer C is incorrect. Neither technician is correct.

Question #87
Answer C is correct. If the air gap is too large, the clutch may slip briefly upon engagement.

Answer A is incorrect. If the compressor clutch coil were defective, the clutch would not engage at all.

Answer B is incorrect. A defective relay will prevent the clutch from engaging.

Answer D is incorrect. A worn clutch bearing may cause a loud clutch engagement but not a slipping.

Question #88
Answer D is correct. Moisture in the refrigerant could cause acid to form when it mixes with the refrigerant, or it could cause ice to form near the expansion device.

Answer A is incorrect. A leaking heater core could cause windshield fogging in the DEFROST mode.

Answer B is incorrect. A clogged evaporator drain could cause windshield fogging in the DEFROST mode.

Answer C is incorrect. An exterior water leak into the air intake plenum could cause windshield fogging in the DEFROST mode.

Question #89
Answer B is correct. The technician should disconnect the batteries before replacing an electronic control panel. This will help prevent a voltage spike when handling the unit. The technician should also ground himself before touching the electronic panel to prevent static electricity from damaging it.

Answer A is incorrect. There is no need to remove control cables from the vehicle before replacing any electronic control panel or device that could have solid-state circuitry.

Answer C is incorrect. Some HVAC electronic control panels can be replaced without disassembling the dash panels.

Answer D is incorrect. Nothing should be applied to the switch contacts.

Question #90
Answer A is correct. The expansion valve is located at the evaporator inlet.

Answer B is incorrect. This is the incorrect location.

Answer C is incorrect. This is the incorrect location.

Answer D is incorrect. This is the incorrect location. The receiver/dryer is located in this location.

Question #91
Answer B is correct. Refrigerant that has been removed and stored is referred to as recovered. Most new A/C machines recycle the refrigerant as the machine recovers the refrigerant from a vehicle.

Answer A is incorrect. Recycled refrigerant has been filtered to remove contaminants and oil.

Answer C is incorrect. Reclaimed refrigerant has also been reprocessed to return it to new product quality.

Answer D is incorrect. Refining is not a process used with refrigerants.

Question #92
Answer B is correct. Only Technician B is correct. Refrigerant suppliers must maintain records about all facilities to which refrigerant is sent. The seller must retain these sales records for three years.

Answer A is incorrect. The purchaser is under no legal obligation to maintain records.

Answer C is incorrect. Only Technician B is correct.

Answer D is incorrect. Only Technician B is correct.

Question #93

Answer B is correct. Only Technician B is correct. The first process is to visually check the components. A technician should also follow the appropriate troubleshooting chart for the proper diagnostic sequence.

Answer A is incorrect. In a strategy-based diagnostic process, a technician must check for visual signs before replacing any parts.

Answer C is incorrect. Only Technician B is correct.

Answer D is incorrect. Only Technician B is correct.

Question #94

Answer A is correct. In high cool mode, the blowers operate at high speed.

Answer B is incorrect. The blowers operate at high speed in high cool.

Answer C is incorrect. The compressor clutch will cycle normally in the high cool mode.

Answer D is incorrect. The blower functions normally in the high cool mode.

Question #95

Answer D is correct. Neither technician is correct. A scan tool is used to recalibrate the HVAC controller with the sensors on some systems. Other systems allow the technician to push a series of buttons on the HVAC control panel and enter diagnostic and calibration mode. The reason Technician B is incorrect is the word "all" in his statement.

Answer A is incorrect. Technician A is incorrect. A/C pressure gauges are used to conduct a performance test of the refrigeration system, not to test the accuracy of the microprocessor system.

Answer B is incorrect. Technician B is incorrect. HVAC system calibration is sometimes not adjustable in the field.

Answer C is incorrect. Neither technician is correct.

Question #96

Answer A is correct. A saline solution will cause corrosion.

Answer B is incorrect. A soft whisk broom can be used to remove debris from the condenser fins.

Answer C is incorrect. Compressed air can be used to remove debris from the condenser fins.

Answer D is incorrect. A soap and water solution can be used to remove debris from the condenser fins.

Question #97

Answer D is correct. After any remaining refrigerant is recovered, the cylinder should be evacuated, marked, and recycled for scrap metal.

Answer A is incorrect. Original refrigerant containers should not be used to store recycled refrigerant.

Answer B is incorrect. If the valve is simply opened, any remaining refrigerant will be vented to the atmosphere.

Answer C is incorrect. There is no reason to introduce oil into the cylinder.

Question #98

Answer B is correct. Only Technician B is correct. This process allows the technician to check the specific coolant flow valve operation.

Answer A is incorrect. Some coolant control valves close when the electric power is removed.

Answer C is incorrect. Only Technician B is correct.

Answer D is incorrect. Only Technician B is correct.

Question #99

Answer C is correct. Both Technicians A and B are correct. Most ATC systems have some type of internal diagnostic routine. You can also display DTCs on the control panel or a scan tool.

Answer A is incorrect. Technician B is also correct.

Answer B is incorrect. Technician A is also correct.

Answer D is incorrect. Both technicians are correct.

Question #100

Answer A is correct. Improper radiator fan operation has nothing to do with the HVAC system. The engine cooling system does not cool the condenser. A defective A/C compressor will not cause the relief valve to trip. If the compressor were defective, it would not be able to develop the high-pressure needed to trip the pressure relief valve. The relief valve typically opens up at approximately 450 to 550 psi.

Answer B is incorrect. A clogged condenser will cause high-pressure.

Answer C is incorrect. A defective high-pressure cutout switch would allow a high-pressure condition.

Answer D is incorrect. A defective compressor will not allow high-pressure to occur.

Question #101

Answer A is correct. The standard coolant-heated heater core type of heater is a forced air convection system.

Answer B is incorrect. An immersion heater is sometimes used to preheat engine coolant.

Answer C is incorrect. Fuel-fired heaters are used as supplemental heat in the northern climates of the United States.

Answer D is incorrect. Electric heaters are not practical for bus HVAC systems.

Question #102

Answer C is correct. The temperature of the lower radiator hose is not a good indicator of an open thermostat.

Answer A is incorrect. Visible coolant flow in the surge tank is not an indication that the thermostat has opened.

Answer B is incorrect. The upper radiator hose gets hot when the thermostat opens and hot coolant begins flowing to the radiator.

Answer D is incorrect. When the thermostat opens, the temperature should stop rising and stabilize in the normal range.

Question #103

Answer A is correct. Refrigerant returns to the compressor as a low-pressure gas. This line is known as the suction line and is very cold on a normally operating system.

Answer B is incorrect. Refrigerant in an operating A/C system is a low-pressure liquid after it passes through the metering device. While in the evaporator, the refrigerant changes from a low-pressure liquid into a low-pressure gas.

Answer C is incorrect. Refrigerant leaves the compressor as a high-pressure gas.

Answer D is incorrect. Refrigerant in the liquid line is a high-pressure liquid. The liquid line connects the condenser to the evaporator. The condenser changes high-pressure vapor into a high-pressure liquid by allowing the refrigerant to release heat and condense into a liquid.

Question #104

Answer B is correct. Only Technician B is correct. Transit bus HVAC systems are manufactured as a separate component. Most have separate controllers that are independent of the bus system.

Answer A is incorrect. Only Technician B is correct.

Answer C is incorrect. Only Technician B is correct.

Answer D is incorrect. Only Technician B is correct

Question #105

Answer D is correct. The easiest way to check the calibration of an ATC system is to measure the interior temperature using a thermometer. Some ATC systems can be calibrated by setting the sensor reading with the temperature recorded on the digital thermometer while in the calibration mode.

Answer A is incorrect. Using an ohmmeter will typically not provide calibration.

Answer B is incorrect. This process only checks that sensor.

Answer C is incorrect. Not all ATC units have such a dial.

Question #106

Answer A is correct. Only Technician A is correct. The unit specification label will identify which refrigerant should be used.

Answer B is incorrect. R-134a should never be vented to the atmosphere.

Answer C is incorrect. Only Technician A is correct.

Answer D is incorrect. Only Technician A is correct.

Question #107

Answer A is correct. Antifreeze raises the boiling point of the coolant, not the conditioner.

Answer B is incorrect. The coolant conditioner cartridge filters particulate from the coolant.

Answer C is incorrect. The coolant conditioner internally lubricates the cooling system.

Answer D is incorrect. The coolant conditioner prevents cavitation corrosion of the cylinder liners.

Question #108

Answer B is correct. Refrigerant leaking from a hose connection will usually contain some refrigeration oil, which quickly collects dirt.

Answer A is incorrect. Excessive pressure in the system would have to be high enough to rupture a hose, and that is unlikely.

Answer C is incorrect. A shaft seal will only cause an oil residue at the front of the compressor.

Answer D is incorrect. Too much oil in the system reduces system efficiency and causes low cooling because the oil will coat the heat exchangers.

Question #109

Answer D is correct. The defroster vent is not part of the A/C system. The evaporator drain is the Most-Likely place to detect refrigerant leaking from the evaporator core.

Answer A is incorrect. Refrigerant leaking from the evaporator core could possibly be detected in the evaporator drains.

Answer B is incorrect. Refrigerant leaks may be detectable above the expansion valve.

Answer C is incorrect. Refrigerant leaking from the evaporator core could possibly be detected at the overhead vents.

Question #110

Answer B is correct. Only Technician B is correct. A technician should follow and perform the diagnostic steps in the truck manufacturer's service manual.

Answer A is incorrect. Replacing the ATC control computer should only take place after following the appropriate diagnostic procedure.

Answer C is incorrect. Only Technician B is correct.

Answer D is incorrect. Only Technician B is correct.

Question #111
Answer D is correct. Low refrigerant charge will cause the clutch to cycle more frequently than normal.

Answer A is incorrect. A defective clutch will not cycle at all.

Answer B is incorrect. A defective control switch will cause the system to be inoperative and have no compressor operation.

Answer C is incorrect. An overcharged system will cause the clutch to remain engaged, and the compressor operation will be louder than normal due to the excess pressures.

Question #112
Answer A is correct. Only Technician A is correct. If the expansion valve is stuck open, the low-side pressure will be high and the compressor will run continuously.

Answer B is incorrect. A defective compressor clutch would cause the pressures to be equal because the compressor would not engage.

Answer C is incorrect. Only Technician A is correct.

Answer D is incorrect. Only Technician A is correct.

Question #113
Answer B is correct. The total weight of the cylinder must not exceed the weight of the cylinder when it is empty plus the maximum rated net weight.

Answer A is incorrect. Cylinder pressure is not an accurate measure of contents level.

Answer C is incorrect. Refrigerant cylinders are not equipped with safety relief valves.

Answer D is incorrect. Shaking the cylinder is not an accurate measure of contents level.

Question #114
Answer B is correct. A technician does not have to recover the refrigerant before removing the control panel.

Answer A is incorrect. It is advisable to remove the negative battery cable before control panel service.

Answer C is incorrect. If the bus contains a PLC system, there is no requirement to wait a specified period after you remove the negative battery cable.

Answer D is incorrect. Self-diagnostic tests may indicate a defective control panel.

Question #115
Answer C is correct. Front-seating a service valve isolates the compressor from the system for service.

Answer A is incorrect. The service valve is front-seated and this blocks the gauge port so no pressure would register on a manifold gauge set. A technician uses the service valve mid-position when servicing the A/C system.

Answer B is incorrect. The system could not operate normally because the passage to the compressor is blocked. Some system pressure may appear at the gauge port.

Answer D is incorrect. Pressure from the compressor is blocked. A technician uses the service valve mid-position for service.

Question #116
Answer D is correct. A blown head gasket is the most likely cause of an internal engine coolant leak.

Answer A is incorrect. The evaporator is not a cooling system component.

Answer B is incorrect. A defective heater valve would have been found during the external inspection.

Answer C is incorrect. A stuck-open thermostat will cause over-cooling of the engine.

Question #117
Answer B is correct. Compressor damage would result if it were operated with the service valve in the front-seated position.

Answer A is incorrect. The back-seated position is the normal operating position.

Answer C is incorrect. The mid-position is used when a manifold gauge set is hooked up.

Answer D is incorrect. The high-side service valve is typically hot when the compressor is running.

Glossary

Absolute Pressure The zero point from which pressure is measured.

Actuator A device that delivers motion in response to an electrical signal.

A/D Converter Abbreviation for Analog-to-Digital Converter.

Additive A substance intended to improve a certain characteristic of a material or fluid.

After-Cooler A heat exchanger often cooled by engine coolant.

Air Conditioning (A/C) The control of air movement, humidity, and temperature by thermodynamics.

Air Dryer Unit that removes moisture.

Ambient Temperature Temperature of the surrounding or prevailing air. The temperature in the service area where testing is taking place.

Amp Abbreviation for ampere.

Ampere The unit for measuring electrical current.

Analog Signal A voltage signal that varies within a given range (from high to low, including all points in between).

Analog-to-Digital Converter (A/D converter) Device that converts analog voltage signals to a digital format.

Analog Volt/Ohmmeter (AVOM) Test meter used for checking voltage and resistance. Analog meters should not be used on solid state circuits.

Anticorrosive A chemical used to protect metal surfaces from corrosion.

Antifreeze A compound, such as ethylene or propylene glycol, added to water to lower its freezing point.

Antirust Agent Additive used with lubricating oils and coolants to prevent rusting of metals.

Armature Rotating component of a (1) starter or other motor, (2) generator, (3) compressor clutch.

ASE Abbreviation for Automotive Service Excellence, a trademark of the National Institute for Automotive Service Excellence.

Atmospheric Pressure The weight of the air at sea level; 14.696 pounds per square inch (psi) or 101.33 kilopascals (kPa).

Axis of Rotation Center line around which a gear or shaft revolves.

Battery Terminal A tapered post or threaded stud on the battery case for connecting cables.

Bimetallic Dissimilar metals joined together that have different bending characteristics when subjected to changes of temperature.

Blade Fuse A type of fuse having two flat, male terminals for insertion in connectors.

Block-Diagnosis Chart A troubleshooting chart that lists symptoms, possible causes, and probable remedies in columns.

Blower Fan A fan that blows air through a ventilation, heater, or air conditioning system.

Boss Bearing race.

British Thermal Unit (Btu) A measure of heat quantity. Amount of heat required to raise the temperature of 1 pound of water 1° F.

Btu Abbreviation for British Thermal Unit.

CAA Abbreviation for Clean Air Act.

Cartridge Fuse Fuse having a strip of low-melting-point metal enclosed in a glass tube. If excessive current flows through the circuit, the fuse element melts, opening the circuit.

Cavitation An erosion condition caused by vapor bubble collapse.

CFC Abbreviation for chlorofluorocarbon.

Check Valve Valve that allows fluid to flow in one direction only.

Chlorofluorocarbon (CFC) A compound used in the production of some refrigerants that damage the ozone layer.

Circuit The complete path of an electrical current, including the generating device. When the path is unbroken, the circuit is closed and current flows. When the circuit continuity is broken, the circuit is open and current flow stops.

Clean Air Act (CAA) Federal regulation, passed in 1992, that resulted in major changes in air conditioning systems.

COE Abbreviation for cab-over-engine.

Coefficient of Friction Amount of friction developed between two objects in physical contact when one is drawn across the other.

Compression Applying pressure to a spring or component attempting to reduce its length in the direction of the compressing force.

Compressor Component of an air conditioning system that compresses low-pressure refrigerant vapor and pumps it through the refrigeration circuit.

Condensation Process by which gas changes state to a liquid.

Condenser Component in an air conditioning system used to cool refrigerant below its boiling point, changing it from a vapor to a liquid.

Conductor Any material that permits the electrical current to flow.

Coolant Liquid that circulates in an engine cooling system, usually a solution of antifreeze and water.

Coolant Heater A component used to aid engine starting and reduce the wear caused by cold starting.

Coolant Hydrometer A tester designed to measure coolant specific gravity and determine the amount of antifreeze protection.

Cooling System Engine system for circulating coolant.

Crankcase The housing within which a crankshaft rotates.

Cycling Repeated on-off action of the air conditioner compressor.

Dampen Reduce oscillation or movement.

Data Links Connection points through which computers communicate with other electronic devices such as control panels, modules, sensors, or other computers.

Deadline Take a vehicle out of service.

Deburring Remove sharp edges from a cut.

Deflection Bending movement as the result of an external force.

Department of Transportation (DOT) A government agency that establishes vehicle standards.

DER Abbreviation for Department of Environmental Resources.

Detergent Additive An additive that helps keep metal surfaces clean and prevents deposits. These additives suspend particles of carbon and oxidized oil in the oil.

Diagnostic Flow Chart Chart that provides a systematic approach to component troubleshooting and repair. Found in service manuals and are vehicle make and model specific.

Dial Caliper A measuring instrument capable of taking inside, outside, depth, and step measurements.

Digital Binary Signal A signal that has two values: on and off.

Digital Diagnostic Reader (DDR) An electronic service tool, usually hand-held or PC based.

Digital Multimeter (DDM) Required to diagnose circuits with solid-state components.

Digital Volt/Ohmmeter (DVOM) Test meter recommended by most manufacturers for use on solid-state circuits.

Diode The simplest semiconductor device, formed by joining P-type semiconductor material with N-type semiconductor material. A diode allows current flow in one direction, but not in the opposite direction.

Dispatch Sheet Form used to track dates when the work is to be completed. Some dispatch sheets follow through each step of the servicing process.

Dog Tracking Off-center tracking of the rear wheels as related to the front wheels.

DOT Abbreviation for Department of Transportation.

Driver's Manual Publication that contains information needed by the driver to understand, operate, and care for the vehicle.

ECM Abbreviation for electronic control module. A controller module that contains the CPU and output generators.

ECU Abbreviation for electronic control unit.

EG Ethylene Glycol Commonly used antifreeze agent.

ELC Extended Life Coolant Premixed coolant solution with high resistance to breakdown.

Electricity The movement of electrons from one place to another.

Electromotive Force (EMF) The force that moves electrons between atoms. Measured in units called volts.

Electronic Control Unit (ECU) Module that processes data and generates outcomes in an electronic circuit.

Electronics The technology of controlling electricity.

Electrons Negatively charged particles orbiting a nucleus.

EMF Abbreviation for electromotive force.

Environmental Protection Agency An agency of the U.S. government with the responsibilities of protecting the environment and enforcing the Clean Air Act (CAA) 1990.

EPA Abbreviation for the Environmental Protection Agency.

Evaporator Component in an air conditioning system used to remove heat from air forced through it.

Fatigue Failures Progressive destruction of a shaft or gear teeth material, usually caused by overloading.

Fault Code A trouble code that is written to computer memory. Can be read by blink-out or DDR.

Federal Motor Vehicle Safety Standard (FMVSS) A federal standard that specifies that all vehicles in the United States be assigned a Vehicle Identification Number (VIN) and adhere to a range of safety standards.

FHWA Abbreviation for Federal Highway Administration.

Fixed-Value Resistor An electrical device designed to have one resistance rating used for controlling voltage.

Flammable Any material that will catch fire or explode.

Flare To spread gradually outward in a bell shape.

FMVSS Abbreviation for Federal Motor Vehicle Safety Standard.

Foot-Pound English unit of measurement for torque.

Franchised Dealership Dealer that has a contract with a manufacturer to sell and service a particular line of vehicles.

Fretting A result of vibration in which the bearing outer race picks up the machining pattern.

Fusible Link A term used for fuse link.

Fuse Link Short length of smaller gauge wire installed in a conductor, usually close to the power source.

GCW Abbreviation for gross combination weight.

Gear A disc-like wheel with external or internal teeth that transmits rotary motion.

Gear Pitch The number of teeth per unit of pitch diameter, an important factor in gear design and operation.

Gross Combination Weight (GCW) Total vehicle weight including payload, fuel, and driver.

Gross Vehicle Weight (GVW) The total weight of a fully equipped vehicle and its payload.

Ground Negatively charged side of an electrical circuit. A ground can be a wire, the negative side of the battery, or the vehicle chassis.

Grounded Circuit A shorted circuit causes current to return to the battery before it has reached its intended destination.

GVW Abbreviation for gross vehicle weight.

Hazardous Materials Any substance that is flammable, explosive, or known to produce adverse health effects in people or the environment.

Heads-Up Display (HUD) A technology used in some vehicles that superimposes data on the driver's field of vision in the windshield. The operator can view the information, which appears to "float" just above the hood. This allows the driver to monitor conditions such as road speed without interrupting his normal view of traffic.

Heater-Control Valve A valve that controls the flow of coolant into the heater core from the engine.

Heat Exchanger Device used to transfer heat, such as a radiator or condenser.

Heavy-Duty Truck A truck that has a GVW of 26,001 pounds or more.

High-Resistant Circuits Circuits that have an increase in circuit resistance, with a corresponding decrease in current.

Hinged-Pawl Switch The simplest type of switch; one that makes or breaks the current of a single conductor.

HUD Abbreviation for heads-up display.

Hydrometer A tester designed to measure the specific gravity of a liquid.

Inboard Toward the centerline of the vehicle.

Inline Fuse A fuse in series with the circuit, usually enclosed in a small plastic fuse holder. Used as a protection device for a portion of a circuit.

Insulator A material, such as rubber or glass, that offers high resistance to the flow of electrons.

Integrated Circuit A solid state component containing diodes, transistors, resistors, capacitors, and other electronic components mounted on a single piece of material.

Jumper Wire A wire used to temporarily bypass a circuit or components for electrical testing. A jumper wire consists of a length of wire with an alligator clip at each end.

Kinetic Energy Energy in motion.

Lateral Runout Wobble or side-to-side movement of a rotating wheel or shaft.

Linkage System of rods and levers used to transmit motion or force.

Low-Maintenance Battery A conventionally vented, lead/acid battery, requiring periodic maintenance.

LTL Abbreviation for less than truckload.

Magnetorque An electromagnetic clutch.

Maintenance-Free Battery A battery that does not require the addition of water during normal service life.

Maintenance Manual A publication containing routine maintenance procedures for vehicle components and systems.

NATEF Abbreviation for National Automotive Technicians Education Foundation.

National Automotive Technicians Education Foundation (NATEF) A foundation having a program of certifying secondary and post secondary automotive and heavy-duty truck training programs.

National Institute for Automotive Service Excellence (ASE) A nonprofit organization that has an established certification program for automotive, heavy-duty truck, auto body repair, and engine machine shop technicians and parts specialists.

Needlenose Pliers Tool with tapered jaws for grasping small parts or for reaching into tight spots. Many needlenose pliers also have cutting edges and a wire stripper.

NIASE Abbreviation for National Institute for Automotive Service Excellence, now abbreviated ASE.

NIOSH Abbreviation for National Institute for Occupation Safety and Health.

NLGI Abbreviation for National Lubricating Grease Institute.

NHTSA Abbreviation for National Highway Traffic Safety Administration.

OEM Abbreviation for original equipment manufacturer.

Ohm A unit of electrical resistance.

Ohm's Law Basic law of electricity stating that in an electrical circuit, current, resistance, and pressure work together in a mathematical relationship.

Open Circuit An electrical circuit whose path has been interrupted or broken either accidentally (a broken wire) or intentionally (a switch turned off).

Oscillation Movement in either fore/aft or side-to-side direction about a pivot point.

OSHA Abbreviation for Occupational Safety and Health Administration.

Out-of-Round Eccentric.

Output Driver Computer output switches in an ECM or ECU. Output drivers are located in the ECU along with the input conditioners, microprocessor, and memory.

Oval Not round; egg-shaped or eccentric.

Overrunning Clutch A clutch mechanism that transmits torque in one direction only.

Parallel Circuit An electrical circuit that provides two or more paths for current flow.

Parts Requisition Form used to order new parts, on which the technician writes the part(s) needed along with the vehicle's VIN.

Payload The weight of the cargo carried by a truck, not including the weight of the body.

PG Propylene Glycol Commonly used antifreeze agent. Should not be mixed with EG.

Pitting Surface irregularities resulting from corrosion.

Polarity The state, either positive or negative, of two electrical poles.

Pounds per Square Inch (psi) A unit of English measure for pressure.

Pressure Force applied to area measured in pounds per square inch (psi) English or kilopascals (kPa) metric.

Pressure Differential The difference in pressure between any two points of a system or a component.

Pressure-Relief Valve Valve located on an air conditioning compressor or pressure vessel that opens if an excessive system pressure is exceeded.

Printed-Circuit Board Electronic circuit board made of thin, nonconductive material onto which conductive metal, such as copper, has been deposited. The metal is then etched by acid, leaving metal lines that form conductive paths for the various circuits on the board. A printed circuit board can hold many complex circuits in a small area.

Programmable Read Only Memory (PROM) A computer memory component that contains program information specific to calibration.

PROM Abbreviation for Programmable Read Only Memory.

psi Abbreviation for pounds per square inch.

RAM Abbreviation for random access memory.

Ram Air Air forced into the engine or passenger compartment by the forward motion of the vehicle.

Random Access Memory (RAM) The memory used during computer operation to store temporary information. The microcomputer can write, read, and erase information from RAM in any order, which is why it is called random.

RCRA Abbreviation for Resource Conservation and Recovery Act.

Reactivity The characteristic of a material that enables it to react violently with air, heat, water, or other materials.

Read-Only Memory (ROM) A type of memory used in microcomputers to store information permanently.

Recall Bulletin A bulletin that pertains to service work or replacement of components in connection with a recall notice.

Reclaim The process used to restore used refrigerants to new product quality. It may include processes available at reprocessing facilities, such as distilling and chemical analysis.

Recovery The process of removing refrigerant from a system and storing it in an external container for further processing or reuse.

Recycling The process of cleaning and removing oil, moisture, and acidity from refrigerant, usually at a repair shop, prior to its reuse.

Reference Voltage The voltage supplied to a sensor by the computer, which acts as a base line voltage; modified by the sensor to act as an input signal. Usually 5 VDC.

Refractometer Refracts light through a liquid to measure specific gravity.

Relay An electric switch that allows a small current to control a much larger one. It consists of a control circuit and a power circuit.

Refrigerant A liquid capable of vaporizing at a low temperature.

Refrigerant Management Center Equipment designed to recover, recycle, and recharge an air conditioning system.

Resistance The opposition to current flow in an electrical circuit.

Revolutions per Minute (rpm) The number of complete turns a member makes in one minute.

Right to Know Law A law passed by the federal government and administered by the Occupational Safety and Health Administration (OSHA) that requires any company that uses or produces hazardous chemicals or substances to inform its employees, customers, and vendors of any potential hazards that may exist in the workplace as a result of using the products.

ROM Abbreviation for read-only memory.

Rotation A term used to describe a gear, shaft, or other device that is turning.

rpm Abbreviation for revolutions per minute.

Rotor The rotating member of an assembly.

Runout The deviation or wobble of a shaft or wheel as it rotates. Runout is measured with a dial indicator.

Screw-Pitch Gauge A gauge used to check the threads per inch of a nut or bolt.

Semiconductor Solid state device that can function as either a conductor or an insulator, depending on how it is biased.

Sensor Electronic device used to monitor conditions for input to a computer.

Series Circuit A circuit with only one path for electron flow.

Series/Parallel Circuit A circuit designed so that both series and parallel combinations exist within the same circuit.

Service Bulletin A publication that provides service tips, field repairs, product improvements, and related information of benefit to service personnel.

Service Manual A manual, published by the manufacturer, which contains service and repair information for vehicle systems and components.

Short Circuit An undesirable connection between two worn or damaged wires.

Solenoid An electromagnet used to perform mechanical work, made with coil wound around an iron tube.

Solid-State Device A device that requires little power to operate and generates little heat.

Solid Wire A single-strand conductor.

Solvent A substance that dissolves other substances.

Spade Fuse A term used for blade fuse.

Spalling Surface fatigue that occurs when chips, scales, or flakes of metal break off due to age rather than wear.

Specialty-Service Shop A shop that specializes in areas such as engine rebuilding, transmission/axle overhauling, brake, air conditioning/heating repairs, and electrical/electronic work.

Specific Gravity The scientific measurement of a liquid based on the ratio of the liquid's mass to an equal volume of distilled water.

Spontaneous Combustion A process by which a combustible material ignites by itself.

Static Balance Balance at rest, or still balance.

Stepped Resistor A resistor designed to have two or more fixed values, by connecting wires to several taps.

Storage Battery Source of direct current electricity for electrical and electronic systems.

Stranded Wire Wire made up of a number of small solid wires, generally twisted together, to form a single conductor.

Structural Member A primary load-bearing portion of the body structure that affects its over-the-road performance or crash-worthiness.

Swage To reduce or taper.

Switch Device used to control on/off and direct current flow in a circuit. A switch can be under the control of the driver or can be self-operating.

System-Protection Valve A valve to protect against an accidental loss of pressure, buildup of excess pressure, or back-flow and reverse airflow.

Tachometer An instrument that indicates rotating speeds, sometimes used to indicate crankshaft rpm.

Tapped Resistor A resistor designed to have two or more fixed values, available by connecting wires to several taps.

Tandem One directly in front of the other and working together.

Throw Number of output circuits of a switch.

Time Guide Reference used for computing compensation payable by the truck manufacturer for repairs or service work to vehicles under warranty, or for other special conditions authorized by the company.

Torque Rotary force.

Toxicity A statement of how poisonous a substance is.

Tractor A motor vehicle that has a fifth wheel and is used for pulling a semitrailer.

Transistor An electronic device produced by joining three sections of semiconductor materials. Used as a switching device.

Tree-Diagnosis Chart A chart used to provide a sequence for what should be inspected or tested when troubleshooting a repair problem.

TTMA Abbreviation for Truck and Trailer Manufacturers Association.

TVW Abbreviation for (1) Total vehicle weight; (2) Towed vehicle weight.

UNEP Abbreviation for United Nations Environment Program. Mandates the eventual phaseout of CFC-based refrigerants.

United Nations Environmental Program (UNEP) A protocol that mandates the eventual phase-out of CFC-based refrigerants.

Vacuum An absence of matter often used to describe any pressure below atmospheric pressure.

Validity List A list supplied by the manufacturer of valid bulletins.

VIN Abbreviation for Vehicle Identification Number.

Viscosity Resistance to fluid sheer. Viscosity describes oil thickness or resistance to flow.

Volt Unit of electromotive force.

Voltage-Generating Sensors These are devices that produce their own input voltage signal.

Watt Measure of electrical power.

Watt's Law Law of electricity used to find the power consumption in an electrical circuit expressed in watts. It states that power equals voltage multiplied by current.

Windings Coil of wire found in a relay or an electrical clutch that provides a magnetic field.

Work The product of a force.

Yield Strength The highest stress a material can stand without permanent deformation or damage, expressed in pounds per square inch (psi).

Zener Diode A variation of the diode, this device functions like a standard diode until a certain voltage is reached. When voltage level reaches this point, the zener diode will allow current to flow in the reverse direction.

Made in the USA
Las Vegas, NV
27 August 2026

53728822R00063